KU-314-442

Understanding Human Engineering

OTHER BOOKS IN THIS SERIES:
Understanding Holography
Michael Wenyon

in preparation:
Understanding Astronomy
Richard Knox
Understanding Futurology
Alan E. Thompson

Vicki Reid
1986

Understanding Human Engineering

John Hammond
C Eng, FIEE, FI Prod E, MBIM

DAVID & CHARLES
Newton Abbot London North Pomfret (Vt) Vancouver

TO BEATRIZ LUCIA

British Library Cataloguing in Publication Data

Hammond, John
Understanding human engineering.
1. Human engineering
I. Title
620.8′2 TA166
ISBN 0–7153–7670–5

Printed in Great Britain
by Biddles Limited, Guildford, Surrey
for David & Charles (Publishers) Limited
Brunel House Newton Abbot Devon

Published in the United States of America
by David & Charles Inc
North Pomfret Vermont 05053 USA

Published in Canada
by Douglas David & Charles Limited
1875 Welch Street North Vancouver BC

Contents

Illustrations

Preface

The aims of this book are to survey a wide field of study and research, specialized in the most part, and to interpret it in simple, direct and concise language. The approach and style derive from lectures which I gave in universities in Colombia and Brazil, and made up at that time for my limited capability in Spanish and Portuguese. I have also called upon experience which I gained devising and making research equipment with my colleagues in the Loughborough University of Technology.

I have tried meticulously to acknowledge all source material; if I have overlooked any, I offer sincere apologies.

Thanks are due to Penrose Anderson for many of the illustrations.

JH

1 Stimulus and Response

Introduction

Ergonomics, from the Greek *ergo*, meaning work, and *nomos*, meaning laws, is the study of man in relation to his work.

The science was born out of World War II. In 1940, the British Government organized the systematic application of science to military operations. The principle was to use a multi-disciplinary approach to solve particular problems, and a number of scientists under Baron P. M. S. Blackett had great success in dealing with a wide variety of projects; e.g., depth charge setting against submarines, optimum size of convoys, tactics in defending convoys, and, later on, the design of equipment. The name given to this team was Blackett's Circus. When war finished two groups of knowledge and techniques emerged, *operations research* and *ergonomics*.

Ergonomics is the term commonly used in Europe, and is equivalent to US terms, such as human engineering, human factors and bio-technology. The main disciplines involved are:

- □ Anatomy and Physiology: Man's structure and functioning
- □ Anthropometry: Body size
- □ Physiological Psychology: Brain and nervous system
- □ Experimental Psychology: Human Behaviour

The central themes running through ergonomics can be expressed in a simplified form by a diagram (fig. 1).

In a man-machine system the various functions can be distinguished as suitable either for machine operation or for manual handling*.

*Fitts and Posner, *Human performance*, Brooks & Cole (1967).

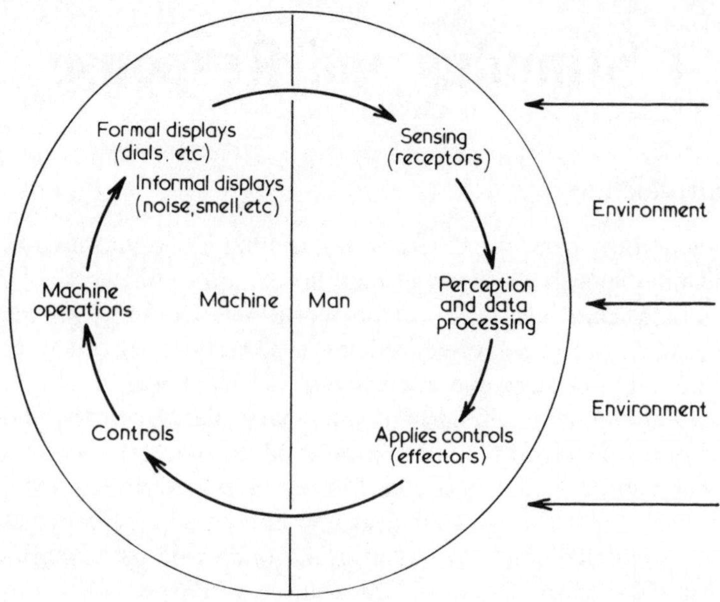

Fig. 1 The man-machine interface: data processing comprises perception, cognition, decision-making, judgement and extrapolation.

Man

- □ senses very small stimuli with limited channel capacity;
- □ has perception;
- □ has a long memory;
- □ improvises and adapts; and
- □ reasons, judges, translates, extrapolates and predicts.

Machine

- □ computes better;
- □ responds more quickly;
- □ applies power*, force, more massively, more smoothly and with more precision;
- □ performs complex simultaneous functions;

*Man, Power/Weight = 0,006 HP/kg; maximum HP = 2,3 one single quick movement.

- ☐ stores a lot of information;
- ☐ does repetitious work;
- ☐ takes Yes–No decisions; and
- ☐ accepts many stimuli at the same time.

The above list takes no account of cost. To replace man's senses of smell or taste or his colour-discrimination by a machine may be impossible, or, if possible, very costly. What the ergonomist aims to do is to ensure that requirements are matched to capabilities, and tasks shared, in accordance with the following factors:

Psychological	*Physiological*	
sensory inputs	posture	force
	functioning	direction
		frequency
		timing

The Human Nervous System

The Neuron

The basis of the nervous system is the neuron, consisting of a nerve cell with nerve fibres which conduct afferent messages into the cell and similar fibres conducting efferent messages out. Neurons are arranged in a network throughout the body; they connect one to another by means of synapses.

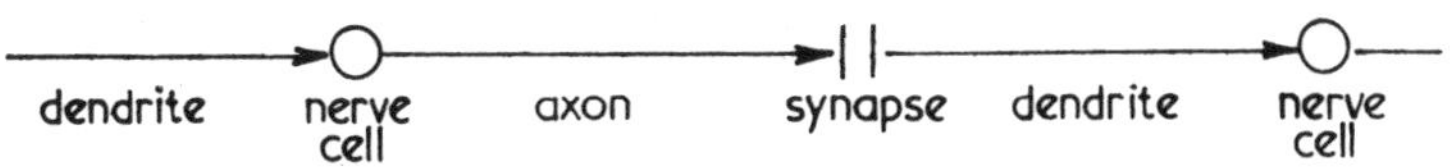

Fig. 2 The neuron channel.

The neuron passes impulses like a train of gunpowder when the first piece of gunpowder is heated beyond its threshold limit. The train is automatically reset after a refractory period during which it cannot be ignited; i.e., when no impulse will be accepted. Impulses travel at about 110m/s; the speed depends partly on the axon diameter. Recovery time before acceptance of a further impulse is 0.001s. The time for an impulse to make a journey through the nervous system will therefore depend upon the route taken, the number of fibres making up the individual neurons, the number of synapses crossed, the length travelled and any perceptual decision-making work done by the brain*.

Organization

The system is organized for two main functions: firstly, to enable man to cope with his environment, and, secondly, to monitor and regulate the internal body. Depending upon the type of activity, the brain pays a greater or lesser part in coordination, evaluation and control. The control is at a minimum in the case of reflex actions.

The *somatic nervous system* is the part enabling man to adjust to his environment; it receives messages from the sense receptors (eyes, ears, nose, taste-buds and skin) and sends messages, in the main, to the effector muscles in all parts of the body.

The part which monitors and regulates the body interior is the *autonic nervous system*, dealing with parasympathetic and sympathetic functions. The parasympathetic functions include, for example, the workings of the heart, digestion and certain glands; they are involuntary and mostly unconscious. The sympathetic system is stimulated by emotions; the resulting actions are mostly involuntary but conscious. Some examples are:

fear	faster heart beat
	panting
	cold sweat
	goose pimples (pilomotor activity)

*Ruch, F. L., *Psychology and Life,* Chapter 17, Scott, Foresman and Co., New York (1964).

guilt	sweaty palms
	sweat generally
greed	salivation
embarrassment	blushing

Reflex action

Reflex action is an automatic reaction to a stimulus (S), and the route taken through the nervous system is known as the reflex arc. When the stimulus is physical the path is invariably through the spinal cord with resulting fast reaction times.

Reflex actions can be caused by both emotional and physical stimuli (we have already seen examples of the former). The classic example of a physical stimulus is the striking of the patellar tendon just below the knee; a pin-prick is another. The brain is aware of the reaction but does not control it and, if the reflex is repeated, the brain becomes progressively less conscious of it.

Conditioned reflex

A reaction which is not natural to a specific stimulus can be developed by association and repetition until it becomes automatic and appears to be a natural reaction. A lot of experimental work with animals has been done, and a lot of the tricks in the circus are based on conditioning.

Many industrial tasks of a repetitive nature become conditioned reflex actions; driving a car, piloting an aircraft, and other similar activities contain many conditioned reflexes. A subliminal stimulus repeated often enough will produce a reflex action. This has interesting possibilities.

Stimulus and reactions

The simple relationship can be represented as in fig. 3.

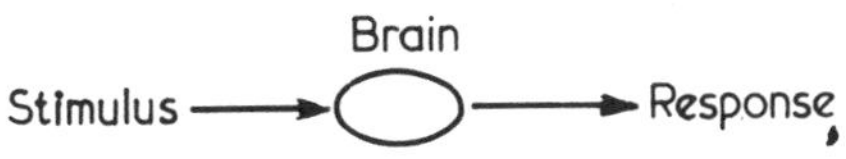

Fig. 3 Stimulus and response.

But there may be one response (R) to several stimuli (S) or several responses to a single stimulus. How these S.R links operate in the brain has been studied by psychologists, in particular by B. F. Skinner. A more complete model can be constructed as in fig. 4.

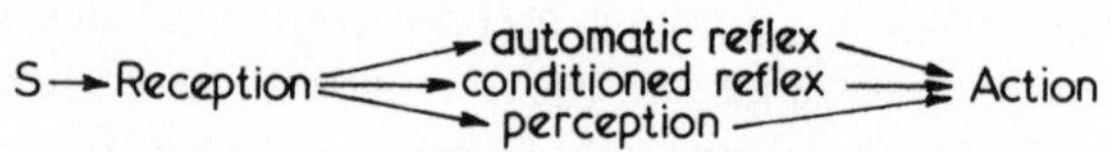

Fig. 4 Stimulus and Action.

Reception

Reception is effected by receptors both on the outside of and within the body; these are called respectively *exteroceptors* and *interoceptors.* The reception is effected to give information which regulates the autonomic functions such as hunger and thirst. *Proprioceptors* are the vestibular and kinaesthetic receptors.

These receptors are biological transducers which convert stimulus energy into electrical impulses in the nerves. These electrical impulses arrive at the brain, where they are filtered and organized, data is placed in storage, data is processed, and decisions are made. Messages are sent as necessary *via* the nervous channels to actuate relevant parts of the physical system of the body.

Perception

Stimuli produce sensations in the sense organs and these, in their turn, set up patterns in the brain. Perception is the process of extracting information from these patterns, mainly by comparing them with patterns stored in the memory. Two people exposed to the same stimulus often perceive differently because perception is a subjective function. There are, of course, discrete stimuli which produce reflex actions—a bang on the nose, a sudden explosive sound, the doctor's rubber hammer just below the knee; however, these are, we hope, not usually found in the working environment. Where a motor activity is the result of a stimulus, the time from the stimulus being applied to the start of the motor activity is defined as the Response Initiation Time (RIT). The processes contained in that time are the perceptual processes. Cognition is used to describe the mental process of identification or interpretation; memory plays a large part in this. Memory is of two kinds, transitory (i.e., short-term) and permanent (i.e., long-term). Learning and experience both involve long-term memory.

The frame of reference and psychological set may quicken the RIT. The RIT is influenced also by:

- the sense channel used;
- the intensity of the stimulus;
- the definiteness of the stimulus;
- the duration of the stimulus;
- the size of the stimulus;
- the number of stimuli;
- the preparedness of the subject; and
- the sex, age, health and arousal of the subject.

Choice of sense channel

The choice of sense channel will depend on the specific nature of the task. A great deal of research has been done on RIT. A simple experimental design has the subject seated at a table with two keys, one by each hand. On perceiving a stimulus his reaction is to press one of the keys. The time between the stimulus being applied and the key being pressed is recorded on each occasion and a measure of RIT is obtained. Table I shows the average RITs recorded in a series of experiments of this nature.

Other things being equal, touch is preferable to sound and sound to vision. For getting attention (e.g., an alarm), sound is the most satisfactory stimulus.

Where the subject expects a stimulus the RIT is appreciably

shorter. This is especially the case when the 'fore' period during which the stimulus is awaited is consistently the same length (as in traffic lights). A time length of between one and four seconds is suitable for the 'fore' period; a state of readiness takes at least one second to build up and cannot be maintained for more than four seconds (as in the starter's drill: 1, 2, 3, *bang!*).

Touch

Tactual sensing is caused by pressure on the skin deforming the skin. If the pressure is maintained, there will be negative adaptation; nothing will be perceived until there is a change in the pressure.

Vibration can be perceived tactually between 10Hz and 8,000Hz with a high ability to discriminate frequences; with practice, it has been shown experimentally, 400Hz and 420Hz can be distinguished. The surface texture of a material is efficiently monitored by a single finger tip. On the other hand, the shape is not well perceived by the contact of a small area of skin. For example, two or more fingers are needed to distinguish the end of a round pencil from that of a hexagonal one*. Some interesting work has been done on the design of equipment controls for the blind using the tactile sense channel†. Similar conditions also apply to controls in night flying, to driving a motor vehicle at night, and to controls which are used when vision is occupied elsewhere.

Channel capacity

Man has the characteristics of a single communication channel of limited capacity. As one datum enters there is a psychological refractory period of about 500 milliseconds†† during

*Fogel, L. J., *Biotechnology*, Section 6, Prentice Hall, New Jersey (1963).

†Rea, G. W., 'A machine shop for blind operators', *The Production Engineer*, London (January, 1968).

††Welford, A. T., 'The psychological refractory period and the timing of high speed performance, a review and a theory', *British Journal of Psychology*, London (1952).

which no more data is accepted. Accordingly, information from different sources and/or by different sensory channels is processed by switching attention from one to another. For this reason drivers are not advised to use the telephone or listen to football commentaries on the radio while at the wheel, although music is usually harmless.

TABLE I
Stimulus and Response Initiation Time (RIT)*

Stimulus and experimental situation	*Average RIT (milli-seconds)*
Expected Touch. (The subject has notice of the approaching stimulus of touch and reacts by pressing a key on the arrival of the touch.)	50
Electric Shock. (The subject presses a key on receiving an electric shock on the hand.)	140
Touch. (The subject presses a key on feeling a touch on the hand.)	175
Sound. (The subject presses a key on hearing a buzzer.)	185
Vision. (The subject presses a key after seeing a light flash on.)	225
Vision with choice. (Subject reacts to two lights. If the right light flashes, the subject presses the right key, and if the left light flashes, the subject presses the left key.)	325

*Adapted from Barnes, R. M., 'Motion and Time Study', *Design and Measurement of Work*, Chapter 15, Table 9, John Wiley, New York (1968).

Strength of stimulus

The threshold limit is set by the amount of noise present in the system (like trying to listen to a single conversation at a cocktail party). This 'background noise' applies to all the senses, and is known as neural noise. Neural noise increases with age; the shortest RITs are to be found with people between 21 and 30 years of age*.

Number of decisions involved

As the number of choices or decisions which are involved in the action to be taken increases, so the RIT becomes longer. Only the data required ought to be presented and symbols should be avoided wherever possible. The more direct and definite the stimulus and the easier the decision process, the better. The load on the operator's memory, whether permanent or transitory, ought also to be kept to the minimum.

Responses and general condition

Recently road accidents and driving in general have been increasingly discussed. Alcohol and tranquillizing drugs adversely affect RITs and motor skills.

The body's diurnal rhythm of arousal is also an important factor. This is measured by the deep body temperature, which changes by approximately 1.5C° from its lowest value in early morning (0400 hours) to the highest values in the afternoon. RITs and motor skills are adversely affected during the times of the lower values. It is safer to set out on a journey by car in the early hours, when the driver's arousal state will be improving, rather than departing in the late afternoon when arousal will be deteriorating. This applies particularly when one is seeking to avoid the periods of heavy traffic.

This body rhythm also gives problems for night-shift workers and for air crews who fly long journeys east or west. For a night-shift worker to be as efficient as he would on a day shift, a twelve-hour phase shift or inversion is necessary. The ability to make this inversion varies with individuals; some take between one and six days or longer and others cannot invert at all. These latter ought not to work on night shifts.

*Bellis, C. J., 'Reaction time and chronological age', *Proc. Soc. Exp. Med. Biol.*, London (1933).

The common week-on-week-off routine is biologically most unsatisfactory. Naturally there are other factors, such as personal preferences and duties*.

Conclusion

One object of this book is to instil in managers an ergonomic approach. The simple man-machine closed loop, shown in fig. 1, can be applied to a wide variety of work systems and work situations.

Many machines found in industry today are nicely styled, with elegant controls. Often, however, they are mounted in such a position that only an operator made in the likeness of a giraffe could effectively handle the controls and at the same time see the work point satisfactorily. And a definite hazard is created when there are unguarded rollers or moving parts with nips in the vicinity of the work point controlled by, say, an inching control which is badly positioned.

There are many examples of stimulus and response in road traffic behaviour. A driver emerging from a minor road onto a busy main road, aiming to join the far stream of traffic, has to pick his time so as to avoid the near stream of passing vehicles. He is subject to stimuli from two different directions. The arrival of an unexpected stimulus, in the form of a pretty girl in a mini-skirt or a dog crossing the road, can overload the driver's channel capacity and create a hazard. In the same way, a driver faced with traffic and road signs in too great a quantity or positioned in unexpected and inconsistent places feels excessive channel loading.

Increasingly today, controllers, both individually and in teams, are required to interpret displays consisting of closed-circuit TV screens, digital read-outs and dials, etc., and control road signals, electric power networks and plant operations of

*Murrell, K. F. H., *Ergonomics*, Chapter 19, Chapman and Hall, London (1965).

all kinds. The ideas contained in this chapter are relevant to layout and operation of such systems. Careful design and layout of the various displays are needed to ensure that each person can see his displays. Linkages between displays and people, and between one person and another, can be arranged to optimize the sensory channels available. Finally, the controls which each man operates must be compatible and consistent with the displays.

2 Light and Vision

Light is that part of the electromagnetic spectrum which stimulates the eye. Its colour depends upon its wavelength (or, equivalently, upon its frequency); the amplitude of the wave determines the intensity of the light. Visible light extends from violet at the shorter wavelengths (higher frequencies) to red at the longer wavelengths (lower frequencies).

Measurement of Light

Assume a point source of light sending out luminous flux in various directions. In each direction the flux will have a certain density; this is called luminous intensity. When the luminous flux falls on a surface, the process of illumination takes place and the flux gives illuminance to the surface. Illuminance is the word introduced in the 1973 IES Code for Interior Lighting to supersede the previous terms, illumination and level of illumination*.

Using SI units, the basic unit for the point source is the candela (cd), which is $1/60 \times$ the luminous intensity of $1 cm^2$ projected area of a black body at the temperature of the solidification of platinum. The unit of luminous flux is the lumen (lm) and that of illuminance is the lux (lx).

When luminous flux falls on a surface, it is reflected or transmitted†—it can also, of course, be partly reflected or partly transmitted; for transmission the surface has to be of a translucent material. Also, part of the flux may be absorbed and converted into heat; this refers particularly to the red (longer wavelength) end of the spectrum.

*Anon., *Code for Interior Lighting*, The Illuminating Engineering Society, London (1973).

†"Transmission" means, essentially, that the light passes through the surface (e.g., of a lens) and is transmitted into the material.

The three factors associated are given below, these relate to non-specular or matt surfaces.

$$\text{Reflection Factor} = \frac{\text{Lux reflected}}{\text{Lux incident}};$$

$$\text{Transmission Factor} = \frac{\text{Lux transmitted}}{\text{Lux incident}};$$

$$\text{Absorption Factor} = \frac{\text{Lux absorbed}}{\text{Lux incident}}.$$

The brightness or luminance of a surface reflecting light is measured by the SI unit, the nit (nt).

Equivalence of Units

1 candela = 1 lumen per steradian. (A steradian is a unit of solid angle; a "model" of a steradian would be a cone of height x whose base had an area of x^2.)

1 lux = 1 lumen per square metre
= 10.76 lumens per square foot.

1 nit = 1 candela per square metre.

Measurement of Colour

Isaac Newton separated white light into the six spectral colours: red, orange, yellow, green, indigo and violet. These are recognized by Man, but the meaning of the terms for the many shades of colour is variable: 'crimson', 'duck-egg blue', 'marigold', etc., mean different colours to different people. For effective work in colour, a classifying and identifying system is needed. This is contained in the Munsell Colour Atlas which consists of over one thousand colour patches in regular steps of hue, value and colour.

- Hue: Red, yellow-green, yellow, green-yellow, green, blue-green, blue, purple-blue, purple, red-purple.
- Value: Darkest (Ideal Black) Lightest (Ideal White).
- Chroma: Neutral grey Maximum saturation or colourfulness.

A shortened colour range based on the Munsell Atlas has been published in BS 4800*. Its application is discussed in *Colouring in Factories* and *The Coordination of Building Colours.*†

The Eye

In discussion of human engineering factors, the features of the eye which are of interest are:

- The cornea: a lens of fixed focal length, like a watch glass, to protect the lens capsule.
- The crystalline lens: an elastic lens in the lens capsule which accommodates to distant and near vision, being flat and thin to view distant objects and bulging and fat for close work.
- The iris: Light enters the eye through the pupil (black) and the iris regulates the flow of light by contracting and lessening the pupil. The iris gives the colour of the eye.
- The retina: a photosensitive mosaic coating of nerve endings consisting of rods and cones, rods on the periphery and cones concentrated on the centre or fovea. The images focussed on the retina are two-dimensional and have a persistence of 20 to 30 milliseconds.
- The foveal area: the most sensitive part of the retina and the spot of most acute vision. To see detail, the object, the centre of the pupil and the fovea must be in alignment. To view an area, the eye samples the scene, jumping or

*Anon., *Paint Colours for Building Purposes*, BSI, London (1972).

†Anon., *Colouring in Factories*, Factory Building Studies 8, HMSO, London (1961); Anon., *The Coordination of Building Colours*, BRE Digest 149, London (1973).

sweeping from point to point: these movements are unconscious. They are more easily made in the horizontal plane. Reading involves a series of pauses and quick jumps.

- The cones: Sensitive to colour, the cones form the content of the fovea; they are insensitive to low levels of illuminance.
- The rods: Sensitive only to low levels of illuminance, the rods are insensitive to colour and are not present in the fovea. They are used for fringe vision and in twilight; they give monochromatic vision only. Fringe vision, awareness without definition, is important in driving a vehicle and much other work.

Illuminance and the Work Place

The Illuminating Engineering Society has published recommended illuminances in lux for a wide variety of tasks and environments; these represent good industrial practice*. Examples are:

General Offices		500
Drawing Offices		750
Bench and Machine Work:	Rough	300
	Medium	500
	Fine	1000
Minute Processes		3000
Stairs		150

To check actual values at a workplace, a light meter, reading in lux, is needed. This contains a photo cell and is similar to the exposure meter used in photography. For a horizontal work surface, the cell must point vertically up; for a wall, out from the wall and at right angles to it.

Ill-lit and badly maintained stairs cause many industrial

*Anon., *IES Code for Interior Lighting* (1977).

accidents. Painting the tread nosings in a light colour value makes for greater safety. In this case, the readings are taken at the treads.

Visual Acuity

Visual acuity describes the capability of the eye to see detail and is measured by the Snellen Chart as used by the optician. If the subject, at 6 metres' distance, can read the seventh line from the top, he has normal vision, described as 6/6 vision. Normal vision is defined as the ability to distinguish detail which subtends at the eye an angle of one minute.

The letter sizes in the chart are dimensioned to subtend one minute of arc at distances reducing with the letter size. Failure above the seventh line indicates subnormal vision and success below it better than normal.

The term 'visual acuity' is used also for discrimination of fine detail in a work task: the many factors have been thoroughly researched by a number of authorities (the reference below contains studies in depth and interestingly described*). Account has to be taken of:

- □ the type of object;
- □ colour hue, chroma and value;
- □ luminance of object;
- □ luminance of background; and
- □ duration of stimulus.

Night Vision

Dark adaptation can take any time from ten to forty minutes or more, depending upon the previous level of illumination. It takes place in a fast phase followed by a slow phase. Its converse, light adaptation, is quicker but not instantaneous, taking up to 2 minutes, again depending upon the level of light to

*Hopkinson, R. G. and Collins, J. B., *The Ergonomics of Lighting*, Macdonald, London (1970).

which one is adapting. Adaptation to dark is quickened by wearing red glasses for a time before entering the dark environment. The dark adapted eye cannot see colours.

Sensitivity to Colour

The eye is not equally stimulated by hues of the same value. The maximum stimulus to the cones occurs at 555 nanometres wavelength—i.e., yellow-green. Hence yellow is used greatly in advertising, the cyclist makes himself conspicuous by wearing yellow, and so forth. Even yellow golfballs have been tried, but golfers are very conservative.

Stereovision

The ability to perceive depth is developed in the growing child by feedback from the muscles, which make the lens adjust and the eyes converge, and also from clues in the visual field. These are:

- □ the known size of common objects (men, cars, trains) in perspective;
- □ atmospheric perspective;
- □ relative motion of near and distant objects; and
- □ the effects of light and shadow.

Perceptual Ambiguity

Visual attention can be distracted by various stimuli. Examples are:

- □ bright lights;
- □ flashing lights;
- □ hues of strong chroma in large areas;
- □ luminance or value contrasts;
- □ specular reflections.

Railway and traffic signals and road-direction signs ought to be well designed and sited with care. Light signals on road vehicles must be limited in size and strength, and advertisements likely to distract drivers must be kept clear of roads. Many accidents have been caused by train drivers, pilots, or aircraft and road vehicle drivers mistaking coloured and other signals due to a confused and ambiguous visual presentation. Often lights and colours are added by new factories built alongside busy roads. Also, the addition of colourful advertisements to a 'safe' scenic situation, as originally planned, can cause accidents.

Visual Illusion

An illusion is mistaken perception. The data may be erroneous and accepted as fact or the data may be correct but interpreted incorrectly. Psychic stress and fatigue contribute to illusions. (The situation in flight has been well studied and described by L. J. Fogel*.)

Autokinetic Illusions

A single fixed point of light against a dark background (e.g. a star) appears to move in a random fashion, this is described as an autokinetic illusion. Moving the eyes and avoiding a steady fixation help to prevent this. Alternately blinking lights as used on aircraft today tend to destroy the illusion.

Visual Perception

The brain organizes the nervous impulses from the retina differently depending upon the individual. The main factors are:

- □ psychological set or frame of reference;
- □ what the subject expects to see;
- □ similar elements grouped together;
- □ completeness (everything being used up); and

*Fogel, L. J., *Biotechnology*, pp151–157, Prentice Hall, New Jersey (1963).

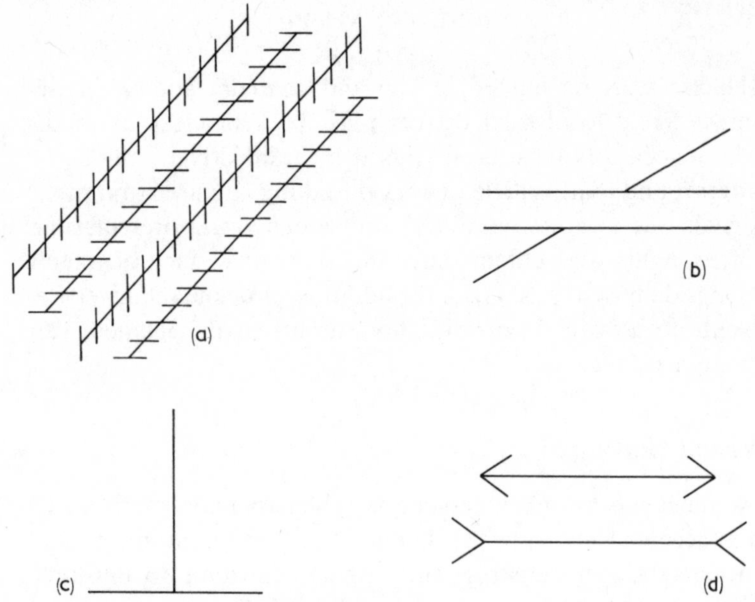

Fig. 5 Some examples of optical illusions.

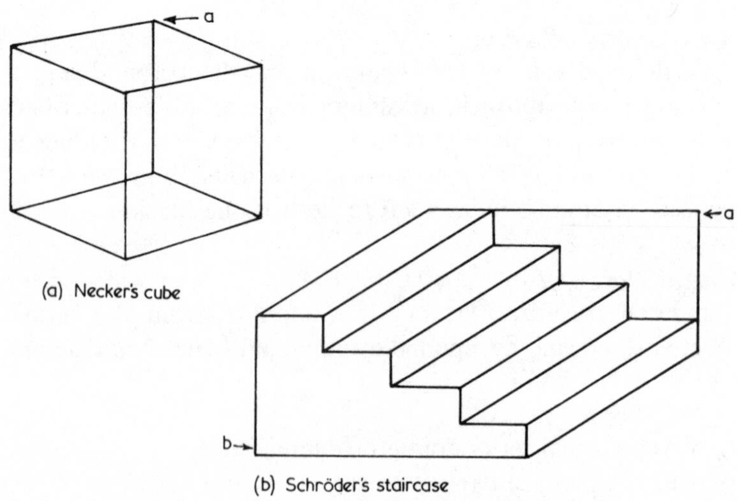

Fig. 6 These three-dimensional objects are represented in two dimensions; depth has to be inferred from clues: (a) a may appear near or far; (b) either a or b may seem to be in the foreground and the staircase may appear to be right or wrong side up.

- ☐ artistic ambiguity—the same picture is deliberately drawn in such a way that it portrays two different pictures, with, usually one being seen closer and the other further away.

I have chosen some examples of objects which either raise doubts or can be interpreted in several ways. These are shown at figs. 5 to 8:

- ☐ fig. 5(a): the long lines are parallel;
- ☐ fig. 5(b): the diagonal line is merely broken—it is the same line;
- ☐ fig. 5(c) and (d): both lines are of equal length;
- ☐ fig. 6(a) and (b): the difficulty of representing three-dimensional objects in two dimensions;
- ☐ fig. 7(a) and (b): these are "impossible" objects;
- ☐ fig. 8: the design can be seen as two Xs, two Vs (one inverted), W over M, or as part of a pattern of diamond shapes.

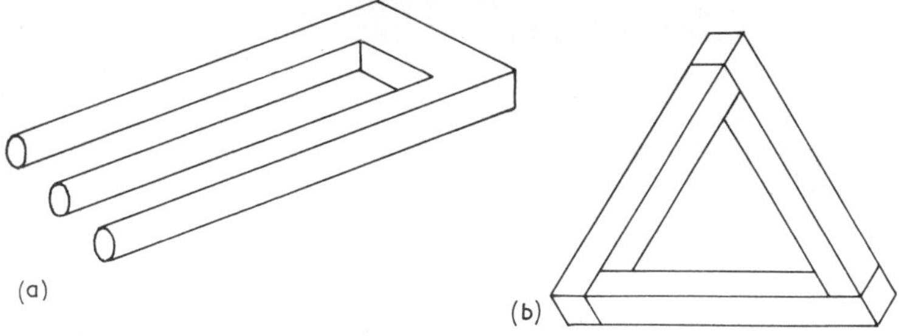

Fig. 7 (a) and (b) are both impossible objects.

Fig. 8 Grouping similar symbols.

Oculo-gyral Illusion

If a person is rotated, gazing at a fixed point against a dark background, the point will seem to move in the direction of rotation while he is being accelerated. When he reaches a constant speed, the apparent movement ceases. Also, when he is being decelerated, the point will appear to move in the opposite direction. The illusion begins to appear at quite small accelerations and decelerations. Aircraft pilots are trained to recognize and allow for these effects.

Successive Stimulation

If the photoreceptors in the retina are stimulated successively in time and in space there are a number of effects caused by the response timelag. I shall deal only with flicker fusion and the phi-phenomenon. (For a thorough treatment, consult Wyburn*.)

If a flashing light is observed, there will be a critical frequency at which the sensation of flicker ceases and the light appears to be continuous. This is known as the critical fusion frequency or critical flicker frequency and varies principally with the light intensity and object size. The range is from 15 to 60 per second.

The phi-phenomenon occurs when two or more stationary stimuli which are near together are received in succession. They are then perceived as in motion. This phenomenon is used in advertising by electric signs and spectacular effects can be produced.

TV and cinema are also examples of the practical use of flicker fusion and the phi-phenomenon.

The cutaneous or touch receptors in the skin react in the same way. A feeling of movement can be generated by successive pinpricks. There is also a similar critical frequency at which repeated pressure stimuli seem continuous: this occurs at approximately 20 per second.

*Wyburn, Pickford and Hirst, *Human Senses and Perception*, Chapter 8, Oliver & Boyd, Edinburgh (1964).

Polarization of Light

Light waves contain electric vibrations in a plane at right angles to their direction. Within this plane these vibrations have no specific direction but change rapidly and at random. However, by passing light rays through a device called a polarizer, the rays emerge vibrating in one specific direction. The incident rays with vibrations in other directions are absorbed, with a resulting loss in the light transmitted. Sunglasses of the polaroid type use this: they are polarized vertically and thus cut out rays polarized horizontally by reflection from the shiny surfaces of motor vehicles, water and polished metalled roads. In this case, the sunglasses act as analyzers, admitting only a specific polarity of light.

The use of these two types of device has been considered as a solution to the problem of vehicle lighting.

Headlamps were fitted with angle 45° polarizers and windscreens with analysers of the same polarity. The driver was able to see the road illuminated by his own headlamps. At the same time, when a vehicle similarly equipped approached him, the light from the approaching vehicle's headlamps was polarized at right angles to the polarization of his windscreen, and hence the glare greatly reduced*. Camber of the road can spoil the geometry of the situation and, for this reason, polarized spectacles are better, allowing the driver to incline his head appropriately. The snags of this scheme were, firstly, that to compensate for the loss of light in polarizers and analyzers, higher-powered lamps were needed with larger capacity generators, and, secondly, that every road-using vehicle had to have the required equipment.

*Anon., *Research on Road Safety*, pp 244–251, HMSO, London (1963).

Psychological Effects of Colour

There are a number of conventions, preferences and psychological associations related to colour. These have become accepted mainly by North Americans and Western Europeans and are not necessarily true in other cultures. The effects are mainly subjective and have not been established scientifically. However, they ought to be considered in design.

Red, Orange, Yellow

- ☐ suggest warmth,
- ☐ stimulate gaiety,
- ☐ appear to advance.

Green, Blue, Blue-Green

- ☐ suggest cold,
- ☐ are less stimulating,
- ☐ appear to recede,
- ☐ are soothing.

- ☐ White makes an object appear large;
- ☐ light colours make an object feel lighter in weight (as with vacuum cleaners);
- ☐ black suggests gloom and boredom;
- ☐ cool colours on side walls and warm colours on end walls make a long narrow room seem wider (as with ships' interiors); a low room seems higher when upper walls and ceiling are of similar light colours;
- ☐ Red and orange indicate danger;
- ☐ Yellow and black diagonally striped indicate moving hazards.

Distractions in the Field of View

These can be caused by:

- ☐ a surrounding lighter in value than the task;
- ☐ a strong chroma hue;
- ☐ an unshaded light source;
- ☐ a reflected light source.

The brightness contrast in the detail of the task may be in values or in hues; e.g., in textiles.

In the case of values, directional lighting may be needed to give more definite shadows. A background target surface or a back-lit translucent screen well diffused may also be useful. In the case of hue contrasts, a hue can be strengthened by incident light of a similar hue to distinguish one coloured detail from another.

Demanding Visual Tasks in Industry

There are many examples of demanding visual tasks in industry; assembly of watches and instruments with fine detail, manufacture of transistors and microcircuits, inspections of surface finishes, electronic instruments, photo film, etc. It is fatiguing to focus both eyes at a close object and this situation must be sought out and steps taken to avoid it. In addition to applying the ideas already described, changing the angle at which the work is presented and the deliberate use of reflected images and shadow are often helpful.

There are also well designed optical aids, from the simple magnifying lens to binocular devices which enhance perception in depth. In inspection, it is easier to make a comparative judgement than an absolute judgement. Three-dimensional colour pictures, colour slides and the portrayal of details graded from acceptable to rejectable can be useful as aids and reference standards.

Colour and Light

The aim of lighting, both natural and artificial, has been stated as an even distribution of light over the whole interior of the building. But flooding the interior with light will not necessarily produce visual efficiency. For the perception of shape and depth in the field of view, a certain amount of shadow is needed. Point sources of light give the most definite shadows and provide directional beams. However, they tend also to be sources of glare, whether or not the individuals concerned are conscious of this. Large window areas and fluorescent tubes, on the other hand, give a diffused effect. Light and colour must be integrated and both visual efficiency and the psychological effects of the colour scheme must be taken into account. Design can be stimulating without being distracting. In the final analysis, what we see is, in effect, a display of luminances; a modern approach to overall design is to study the ratios and pattern of luminances and to adjust and balance them to obtain the visual effects and quality desired*.

Conclusions

Many reasons are given for the continual increase of traffic accidents in both quantity and severity. A great many accidents are the result of perceptual ambiguity. Pile-ups on the motorway often start from one driver's misjudgement of the distance and speed of the vehicle in front of him: clues may, perhaps, have been reduced by darkness and fog to the image of two red lights and nothing more; their brightness, distance apart and steadiness in the field of view have to be interpreted to give distance and speed values. In this situation, it can be difficult to distinguish between a stationary vehicle and one in motion.

Regarding designing in light and colour, I have discussed some of the factors and referred to the main authorities, the Illuminating Engineering Society, the Electricity Council and the Lighting Industry Federation Ltd. Attention to the general visual environment and to the design and layout of the individual work-place and task in terms of colour and light are worthy objectives and pay for themselves in enhanced worker satisfaction as well as greater efficiency.

Tables II, III and IV summarize some of the ideas expressed in this chapter and form useful *aides-memoire.*

*Durant, D. W. (editor), *Interior Lighting Design*, Lighting Industry Federation Ltd. and the Elctricity Council, London (1973) is an authoritative manual of design and well worth consultation.

TABLE II
Light in Action

Desirable	*Undesirable*
A good general distribution of light over the whole area.	Shadows created by large equipment, parts of building, etc.
Plenty of reflected light from ceilings, floors and walls.	Fenestration which gives unidirectional light.
Windows set horizontally or angled not more than 45° to the horizontal.	Glare* from: ☐ light sources in the field of view; ☐ brightness contrasts; ☐ reflected light sources; and ☐ excessive penetration of sunlight.
Needs for efficient maintenance and cleaning of windows and light fittings to be considered and provided.	
Desirable features in light fittings: ☐ scouring or self-cleaning effect; ☐ minimum cut-off 45° to horizontal; and ☐ splash some light on ceiling.	Monotony and visual boredom caused by a uniformly bright field of view with 100% fluorescent lighting.
	Flicker and stroboscopic effects due to fluorescent lamps incorrectly wired.

*An exception is the requirement for an attention attracting signal. In this case, high contrast is needed; e.g., orange or red on a black background.

References: *The Lighting of Factories*, HMSO, London (1959). *Colouring in Factories*, HMSO, London (1961).

TABLE III
Colour in Action

Desirable	*Undesirable*
Harmony: few hues; each hue distinct and physically separate; one dominant hue; complementary hues.	Large areas of strong chroma. *Unpleasing combinations:* hues of strong chroma and equal value adjacent in the colour circle; e.g., red and orange; violet and blue.
Balance: warm and cool colours in balance; towards warm except for hot processes or in hot climates.	

Desirable	*Undesirable*
Stimulation: some bold colour in surrounds and ends of passages.	Dazzle and visual aberrations due to certain conditions; e.g., strong chroma red with neutral grey; strong chroma red with strong chroma blue; strong chroma red with strong chroma green; all of same value.
Reflection: selection of hues and values on walls and main surfaces to reflect and distribute light; (typical Munsell values: Ceilings 9; Walls 8; Floors 6).	

Reference: *Colouring in Factories*, HMSO, London (1961).

TABLE IV
The Visual Task, Light and Colour

Desirable	*Undesirable*
Illumination in lux consistent with detail size and detail contrast.	Unshaded light sources; spectacular reflections; strong hues (e.g., colour coded pipes) — In the line of sight.
Careful ergonomic design for acuities less than 6 minutes of arc, taking account of nature of task, contrasts, quality of illumination, etc.	Light directionless or so arranged as to destroy shadows in the detail of the task which are useful to the worker.
Contrast in detail high in value and distinct in hue; background and surrounds darker in value and without distractions.	Coloured light (e.g., fluorescent lamps) which changes human appearance and coloured surfaces.
Colour of light consistent with task. Important area of displays to be: subtending horizontally not more than 65°; not more than 10° above the eye and 45° below the eye.	Letters, numerals and signs of high-value hue on low-value background (e.g., white on black) except for the dark-adapted eye.
Height of letters and numerals:	
Display panels Artificial light 0.005 × Reading Distance;	
Road signs Daylight 0.0025 × Reading Distance.	

Reference: McCormick, E., *Human Engineering*, McGraw Hill, New York (1957).

3 Sound and Hearing

Introduction

Sound is a form of energy propagated molecule-to-molecule as a pressure wave in air or any other elastic medium. Its speed in air at 20°C is 344m/s (about 1200kph) and, if produced by a point source, the wave form is spherical, its intensity diminishing with distance according to the inverse square law. It is reflected at an interface with a loss of energy dependent upon the nature of the material.

Transmission of sound through a solid material is caused by setting up, in the material, vibration which is re-radiated on the other side. In case of a partition the weight per unit area is a significant factor: a dense material is the more effective reducer of such transmission of sound. Sound, as perceived by the listener, is of both objective and subjective nature.

As a waveform it is defined by its frequency and amplitude. The frequency is the tone or pitch and the amplitude the intensity or loudness. A wide range of frequencies can be sounded in air but only those in the range 16Hz to 20,000Hz stimulate the ear.

Ultrasonic frequencies have recently become important owing to their many industrial applications: welding dissimilar metals and plastics, machining ceramics and glass, nondestructive testing of materials and many other uses. (The term supersonic is restricted to the speeds exceeding 344m/s.)

Units

The system of units is complicated by the fact that a sound which is uncomfortable to the ear has an energy level 10^{12} times that which is first heard by the human ear, and also by the need for subjective measurements. The following are the units most commonly used*.

Bel and decibel (dB) are dimensionless objective units which measure the sound pressure level (SPL) on a logarithmic scale with reference to a threshold pressure on the ear of 20 micro-neons per square metre. This would be a sound just heard by a young person with normal hearing, and is defined as 0db. Sound pressure levels are measured with a sound level meter giving direct readings in dB and having a stated response to specific frequency bands, or 'weighting' as it is termed. Typical SPLs would be: noise in a large office, 50dB; pneumatic hammers used on roads or a train passing, 100dB.

Sound levels above 120dB are painful to most listeners and no one ought to be subjected to levels above 135dB, no matter how short the duration.

Subjective Units

The **mel** is a unit of pitch. A simple tone of 1,000Hz frequency, which is 40dB above a listener's threshold, produces a pitch of 1,000 mels.

The pitch of a sound judged to be n times that of a 10^3 mel tone is $n \times 10^3$ mels.

The **phon** is a unit of loudness level. Perception of loudness depends upon the sound pressure level, the frequency spectrum and the attitude and emotions of the listener. The phon is derived from psycho-acoustical experiments on a number of subjects and is their judgement of equal loudness relevant to a reference sound in dB at a tone of 1,000Hz. *Handbook of Noise Measurement* (see footnote, p38) contains a discussion of psycho-acoustical measurement.

The **sone** is a unit of proportional loudness. For engineering purposes, a unit was needed which corresponded better to the

**Final Report of the Committee on the Problem of Noise*, Chapter 1 and Appendix II, HMSO, London (1968). Peterson, A. P. G., and Gross, E. E., *Handbook of Noise Measurement*, General Radio Co., West Concord, Mass., USA (1967).

relative sensation of the listener. This was the sone and was standardized internationally. The relationship between sones and phons is given below. A simple tone of 1,000Hz and 40dB above a listener's hearing threshold produces 1 sone. The loudness of any sound judged to be n times that of the 1 sone tone is n sones.

Relationship between Sones and Phons

$$S = 2^{(P-40)/10}$$

where S is in sones and P in phons.

Example: If $P = 72$, then $S = 2^{(72-40)/10} = 2^{3.2}$. Taking logarithms, $\log S = 3.2 \log 2 = 3.2 \times .301 = .9632$, and so $S = 9.2$ sones.

The extent of the scale from a barely discernible sound to a painful one would be in phons 50 to 100 (twice), but in sones 2 to 64, a more representative and true scale.

Reflection of Sound

When sound strikes a surface, some of it is absorbed, some is transmitted and the rest is reflected. In the case of a plane surface, if the area is dimensionally greater than the wavelength, the angle of reflection will equal the angle of incidence. A room containing little absorbent material is acoustically 'live' and a listener will receive from a source in the room sound reflected from walls, floor, ceiling, etc., as well as directly from the source itself. The domestic bathroom has these characteristics, and accordingly makes an indifferent male singer sound like a rich operatic baritone.

A 'reverberation room', especially designed and with minimum absorption, is used to find the absorption values of materials and to measure sound power in instances.

The opposite of this room is an 'an-echoic room', which has

maximum absorption and is as acoustically dead as possible. This effectively provides a free field, uniform and with no interference from reflected sound. In this environment we can measure acoustic power and the acoustic directivity pattern of a piece of equipment before it is installed, to meet specifications.

Modern Man a Victim of Noise

In 1960 the British government appointed a committee to examine and report on the problem of noise. Their final report* describes their findings. Chapters are included on motor vehicles and aircraft. The *London Noise Survey*† covers the objective urban problem and the subjective effects of noise on urban dwellers. A difficulty is to measure the irritation and disturbance which noise creates. Another is that people become tolerant of noise but nonetheless suffer damage to hearing which may be either temporary or permanent. In the case of permanent damage, the matter is also complicated by the spectrum of noise being considered, and the fact that it occurs over a long period during which hearing also deteriorates from the simple effects of age.

We shall deal with the problem of noise in industry in detail in the next chapter.

Damage risk levels

This most complex subject is thoroughly discussed by Murrell††. Guidelines are:

1). For any noise reaching the operators' (or other workers') ears attaining the SPLs shown in Table V 8 hours a day, 5

*Murrell, K. F. H., *Ergonomics*, Chapter 13, Chapman and Hall, London (1965).

†*London Noise Survey*, Dd 135642, HMSO, London (1968).

††See Footnote, p38.

days a week, protection ought to be provided or other measures taken.

2). No unprotected ear ought at any time, for no matter how short a period, be exposed to a noise of SPL greater than 135dB.

3). Audiometric examinations of workers exposed to noise ought to be made periodically. (The audiogram is a chart of hearing loss against frequency.)

TABLE V
Damage risks

Frequency Band (Hz)	SPL (dB)
37.5 – 150	100
150 – 300	90
300 – 600	85
600 – 1,200	85
1,200 – 2,400	80
2,400 – 4,800	80

TABLE VI
Speech Interference and Communication

SIL (dB)	*Voice level and distance for word intelligibility*	*Nature of communication possible*
45	Normal voice at 3 m	Relaxed conversation
55	Normal voice at 0.9 m Raised voice at 1.8 m Very loud voice at 3.6 m	Continuous communication in work areas
65	Raised voice at 0.6 m Very loud voice at 1.2 m Shouting at 2.4 m	Intermittent communication
75	Very loud voice at 0.3 m Shouting at 0.6 to 0.9 m	Minimal communication
85	Shouting at 0.3 m	Minimal communication

Masking and speech interference

We are used to experiencing the masking of a sound by a louder sound. Sounds close in frequency and pure in tone mask each other with ease, tones of different frequencies less so. Broad-band noise masks other broad-band noise, but a pure tone may be heard even if it is lower in pressure level than the masking sound; e.g., a 1,000Hz tone can be heard 10dB down on a masking sound with a 600–1,200Hz spread.

In a work environment the background noise likely to mask speech is normally broad-band random noise. Accordingly, in order to measure the level at which speech is masked, analysis has been standardized on the arithmetical averages of the noise levels in the three octave bands 600–1,200Hz, 1,200–2,400Hz and 2,400–4,800Hz. This then gives the Speech Interference Level (SIL). Table VI shows conditions for communication by speech at different SIL*.

*Table VI is adapted from Aldersley-Williams, A. G., *Noise in Factories*, Factory Building Studies No. 6, Her Majesty's Stationery Office, London (1960).

4 Noise and Industry

I am indebted to the Building Research Station, UK, and their report on Noise in Factories* for many of the ideas and techniques which I describe in this chapter.

Sources of Noise and Vibration

In the case of industry in general, we can place equipment which generates vibration and noise into two classes:

- Heavy, typified by the noise of an impact of short duration and transient nature; e.g., hammers, presses, guillotines, forges.
- Light, producing noise of less amplitude but higher frequencies; e.g., electric motors, ventilators, pneumatic nut runners, high-speed pneumatic drills, sanders, polishers.

The descriptions heavy and light do not apply specifically to the weight of the equipment, alone, but to this coupled with the type of vibration and the sounds produced.

Heavy equipment

Apart from the problem of airborne noise, heavy equipment readily transmits vibrations with little attenuation along the floor, steel framework, pipes, etc. If such equipment cannot be isolated in a distant part of the factory, the careful design of the foundations will be important. The inclusion of resilient materials in mountings will often reduce transmitted noise.

*Aldersley-Williams, A. G., *Noise in Factories*, Factory Building Studies No. 6, HMSO, London (1960).

Light equipment
With light equipment there is certainly the problem of airborne noise, but there may also be vibrations conducted by the floor and framework—i.e., solid-borne vibration—as in the case of heavy equipment. Being of high frequencies, these can cause considerable distress.

Control of Vibration and Sound

The first step in control is to measure the SPL and spectrum of the source and investigate its directional pattern. Depending upon the nature of the source and the acoustic environment, the following actions will be appropriate either separately or in combination.

Reduction of sympathetic or resonant vibrations
Sound is radiated by any surface in vibration in the same way as by a diaphram in a loudspeaker. This is a common problem in motor-vehicle design in particular, evidenced by sympathetic vibrations of the car body known as 'body boom'. These in turn are excited and sustained by solid-borne vibrations of moving parts.

Doors and partitions of light construction can pass noise in a like manner (in this case airborne noise). The sound reduction varies directly as the weight / unit area of the structure and the frequency. The structure ought to be as airtight as possible and attention ought to be given to the transmission parts on the perimeter for solid-borne vibrations. Double glazing in walls has a maximum reduction factor of 40dB: the space between the glazing must be between 10cm and 20cm for practical effectiveness.

Introduction of a hiatus or resilient material in the path of solid-borne vibrations
The most common application is to the mounting of a piece of equipment like an electric motor. The reduction in trans-

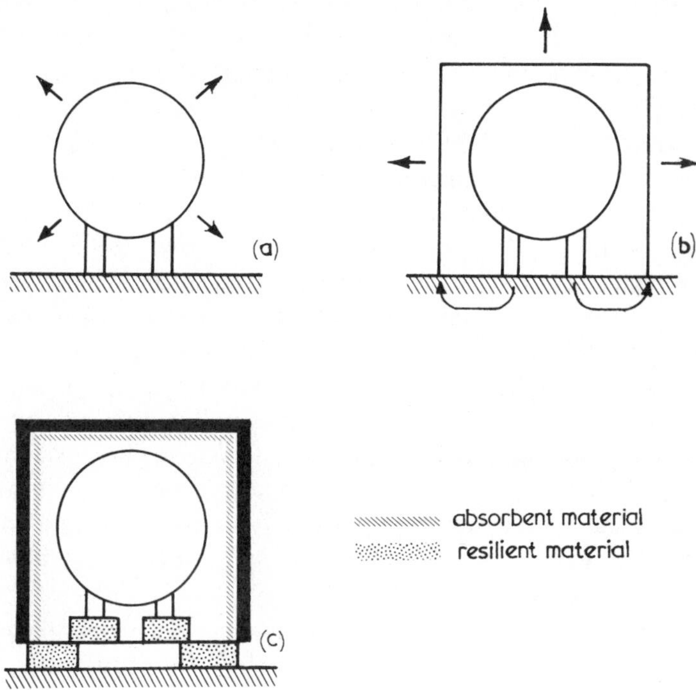

Fig. 9 Reduction of solid-borne and airborne noise. In (a) airborne noise is primary while solid-borne noise is minimal. In (b), with the addition of a rigid sealed enclosure, vibration is sent through supports to the floor, thence to the enclosure and thence to the outside; in (c) the addition of absorbent material and vibration-isolation mounts of resilient material produces definite noise reduction, although use of unsuitable resilient material can increase vibrations at certain frequencies.

mitted vibration or noise depends upon the ratio of the forcing frequency of the equipment to the natural frequency of the resiliently mounted system. Sometimes vibration-producing machines can be isolated by mounting that part of the floor which supports them on resilient mountings.

However, the efficiency of resilient mounting is often lessened by allowing a solid-borne short-circuit to remain; e.g., a fuel pipe, an exhaust pipe, or a control. In many cases, a combination of measures may be necessary—for example, the use of resilient mountings, the enclosure of the source, and the lining of the enclosure with absorbent material. Figure 9 illustrates some methods of noise reduction.

From the electric motor shown at 9(a) noise radiates directly to the air and solid-borne noise also travels along the floor. The first effect of enclosing the motor—as shown in 9(b)—will be a high level of noise energy inside the enclosure from reflections. If the material has a low weight per unit area there will be a considerable transmission to the exterior. The stiffness of the five sides of the enclosure is also significant, as sympathetic vibrations will be set up and each panel will vibrate and send out noise. The mountings supporting the motor are securely fixed to the floor, to which they form an efficient transmission path; some vibration travels back to the enclosure and the remainder carries along the floor. In fact, little, if any, noise reduction has been achieved.

At 9(c), however, a more rigid and massive material has been used to enclose the motor. Secondly, absorbent material has been attached to the interior surfaces. The third measure is the insertion of resilient pads between the motor and the enclosure and between the enclosure and the floor. These three measures have:

- reduced transmission through the panels;
- reduced internal sound energy; and
- reduced transmission through the mountings.

The resilient material must be carefully selected: with an unsuitable material it is possible that certain frequencies in the sound spectrum would be actually increased in terms of sound pressure level.

Reduction of reflected noise

The direction pattern of a source and the main reflected waves can indicate the use of absorbent material and alteration in the position of the source. Early work on absorption was done by Wallace Sabine, who established the unit of absorption, the sabin, as the absorption equivalent to $1ft^2$ ($0.93m^2$) of a perfectly absorbent surface* which, of course, does not reflect any sound. An open window of $1ft^2$ has the same effect, so the sabin is also known as the 'Open Window' unit. The metre-sabin, where the imaginary surface has an area of $1m^2$, is also used.

The coefficient of absorption is given by

$$\frac{\text{Sound energy absorbed at a surface}}{\text{Sound energy incident upon that surface}} .$$

Measurements are made in a room and an estimate of the reverberation time can be made from Sabine's formula:

$$T = \frac{0.049V}{S_1\,\alpha_1 + S_2\,\alpha_2 + \ldots\ldots\ldots\ldots S_n\,\alpha_n} ,$$

where T = reverberation time in seconds,
V = volume of room in ft^3,
S = surface areas in ft^2,
α = absorption coefficients of the various surfaces in sabins.

For metric units the constant 0.049 becomes $0.049 \times 3.28 = 0.16$.

The reverberation time is defined as the time for the SPL (RMS value) originally in a steady state to decrease 60dB after the source of the sound is stopped. The slope of decline, which can be traced by means of an oscillograph or graphic recorder, is known as the decay rate and measured in dB/seconds.

*Sabine, W. C., *Collected papers on acoustics* (1922).

Absorption coefficients are quoted for frequencies in seven-octave bands but, for most calculations, three suffice; e.g.,

	125 Hz	500 Hz	2000 Hz
Plaster	0.02	0.02	0.04
Concrete	0.02	0.02	0.05
Glass, 3mm thick	0.35	0.18	0.07
Acoustic tiles	0.40	0.70	0.90
Air per m^3	—	—	0.007

Although treating the walls and ceiling of a factory with an absorbent material makes no difference to the direct noise, the reflected sounds can be considerably reduced and, where there is a high level of random noise, the environment be rendered less acoustically confusing.

An example of the application of Sabine's formula is given in the Appendix.

Investigation of the use of earplugs or ear muffs

Earplugs which fit well in the ear canal can be effective in certain cases. Sound conduction by the bone structure of the head is, of course, not reduced in any way. Earplugs can also be selective in terms of the frequencies which they exclude. Ear muffs with liquid seals in the form of ring-shaped cushions are also very effective.

Measurements of Sound

In the early days of electronics, when the cathode-ray oscilloscope was first developed, a party piece for visiting dignitaries was to show them the waveforms of their voices on the screen. The trace jumped up and down in an apparently random fashion; a clear waveform would be traced only when a pure or nearly pure tone was presented to the microphone.

If the jumble of sounds typical of modern noise forms the input to a meter which simply indicates sound pressure by a pointer moving over a scale, the pointer will invariably flutter up and down so that the best that one can do would be to estimate the average value. But damping down the flutter and estimating the average value are no problems to the electronics engineer. In practice, then, the measurement system will include a microphone, an attenuator, an amplifier and a network of circuits to derive the effective (or root mean square) value over a given time. This will be connected with the indicating unit, which can be a pointer on a scale, a pen recording on a roll of paper or a print-out. The microphone will be receiving a band of frequencies of sound; however, the sensitivity of the human ear varies for different intensities and frequencies. If the meter can be designed to respond appropriately to different frequencies, the resulting values become, in effect, subjective and indicate how people would judge the sound themselves.

The networks incorporated in the meter for this purpose are known as weighting networks; they are standardized internationally and are designated A, B and C. The A scale is the most commonly used, and sound levels so measured are distinguished with the suffix A, as in dBA (dBB and dBC are, of course, also used).

Sound measurements and related social surveys are described and discussed in the *Final Report of the Committee on the Problem of Noise* (see footnote, p38) and in *London Noise Survey* (see footnote, p40).

Speech and Communication

There are basically three requirements for efficient communication by speech in lecture rooms, conference halls, etc. These are:

- ☐ that the speaker's face can be clearly seen;
- ☐ that the speech interference level (SIL) is not exceeded as a result of exterior noise; and
- ☐ that the acoustic environment is not too live or reverberant.

In the theatre, it is usual for the actor to speak important lines downstage and straight out to the audience; also he will have his features well lit. If he faces in the upstage direction, on the other hand, with his back to the audience, he has to speak more slowly and enunciate more clearly. Accordingly, in the lecture theatre, good lighting on the lecturer's face is important.

Regarding SILs (see p42), reasonable values for lecture theatres would be not more than 40dB.

Speech and Reverberant Sound

In a lecture theatre, speech will arrive at each member of the audience both directly and reflected from the walls, ceiling and other reflecting surfaces. The reflected speech is known as reverberant sound. The reverberant sound will also have varying spectra since the absorbent surfaces have different coefficients of absorption relative to frequency. An approximate maximum difference between direct and reflected paths can be taken as 18m: there will be difficulties due to interference above this limit*. The optimum distance from speaker to listener has been quoted by some authorities as about 4.5m; beyond 6m the direct sound falls off. However, the reverberant sound from a first reflection can be used provided that the two paths do not differ by more than 18m.

For a satisfactory design, this limit can well be reduced and the preferred reflecting surface used can be the ceiling. The procedure is to measure or estimate the reverberation time with a two-thirds audience present and then to add absorbent material until the time approximates the reverberation time required.

Balancing the reflected sound

In order to satisfy the aim of keeping the first reflections strong and absorbing the multi-reflected sounds, the order in which

*Parkin, P. H., and Humphreys, H. R., *Acoustics, Noise and Buildings*, Chapter 3, Faber and Faber, London (1958).

the absorbent materials are placed ought to follow: first, the back wall; second, the edges of the ceiling; and, third, the side walls.

One can also use sounding boards which reflect more strongly a certain range of frequencies and at the same time reduce the path of certain first reflected sounds. For speech, a high level of reflection is desirable between 200 and 6,000Hz although the spectrum is from 100–10,000Hz. Music on the other hand has a spectrum between 60Hz and 15,000Hz with some important musical sounds in the high-frequency bands.

5 Body Balance and the Senses

Balance and the Position of the Body

That people are able to stand erect, and even balance on a moving surface, supported by the relatively small area of two feet is due to the proprioreceptors. These consist of the vestibular mechanism and the kinaesthetic receptors.

The *vestibular mechanism* is contained in the inner ear, where there are three semicircular channels, roughly at right angles to each other. The pressure of the fluid in these channels changes with the position of the body and these changes are monitored and interpreted by the brain. Dizziness, motion sickness and weightlessness or zero-g sickness are all related to the vestibular mechanism.

The *kinaesthetic receptors* are attached to the muscles, tendons and joints of the body and send signals to the brain of stretch and pressure; in this way the positions of the various parts, and to what positions they have moved in answer to directions from the brain, are known. They form perfect closed loop feedback systems.

Kinaesthesis enables us to do many things in the dark. It features strongly in athletics and certain displays of skill; for example, driving a golf ball blindfold. It is important in many industrial crafts and in everyday activities such as driving and bicycling.

The Skin Senses

If four different types of stimulus—touch, warmth, cold, and pain—are applied to the body, four different maps can be drawn indicating different areas of sensitivity. It is thus shown that there are four distinct skin senses.

We are concerned mainly with touch. The most sensitive areas are the lips, the tip of the tongue, the fingers and the hands. The stimuli that produce the feeling of touch are the movement of body hair and the deformation of the skin. When a stimulus is maintained, the sense of touch quickly adapts.

In industry the hand, with its ridges and grooves which increase its sensitivity, is the main part of the body used. In many work situations, touch can be used when the perceptual loading through sight and hearing is already heavy. At night, or when the eyes have to concentrate on a target area, touch can be of considerable importance. Perception of shape by single touch alone is difficult: the use of the fingers is needed for reliable perception. Certain shapes and extremities are at once identifiable*.

The sensing of the temperature of objects depends upon the temperature difference between the skin and the object's surface and the conductivity of the interface. Where sight is not used, sensing temperature can provide a clue for the perception of the material of an object.

Taste and Smell

The receptors for taste respond to materials in solution; that is, liquids. Those for smell respond to volatile substances. The former consist of taste buds, lying mainly on the top and sides of the tongue; the olfactory receptors are situated slightly to the right and left of the upper part of the nasal cavity. In this way they are not stimulated continuously by normal breathing, but can be deliberately stimulated by sniffing.

In tasting, smell plays an important part, often at an unconscious level; the 'taste' of an onion consists almost

*See Kellermann, van Wely and Willems, *Vademecum, Ergonomics in Industry*, Chapter 6, Philips Technical Library, Eindhoven (1963).

entirely of its smell. Taste is easily masked by smell; for example, cigarette or cigar smoke. At present, these two senses cannot be effectively replaced by the machine and both are very sensitive (especially smell). They play an essential part in the foods, drinks and tobacco industries; and smell often acts as a warning of overheating motor car engines, gas leaks and similar emergencies.

6 Mechanical and Physiological Work

When a force moves its point of application a certain distance, work is done and the quantity of work is measured by the product of the force and the distance. In SI units the force is expressed in Newtons (N) and the distance in metres (m): the mechanical work done is therefore expressed in Nm.

Physiological work is not identical with mechanical work. Mechanical work takes no account of the posture of the body, known as static or postural work, and the work done in gripping tools, workpieces and so on. A load is also placed on people by the environment, temperature, humidity and noise.

At this stage I shall not consider the many jobs involving what we loosely call mental and nervous energy. High perceptual alertness, interpretation of signals, decision-making and judgment are some of the many characteristics which are difficult to define in specific terms, let alone measure.

Physiological Work

Muscular activity changes the following functions:

- heart rate,
- blood pressure,
- cardiac output (litres/minute),
- chemical composition of the blood and urine,
- body temperature,
- perspiration rate,
- pulmonary ventilation (litres/minute) and
- oxygen consumption by the muscles.

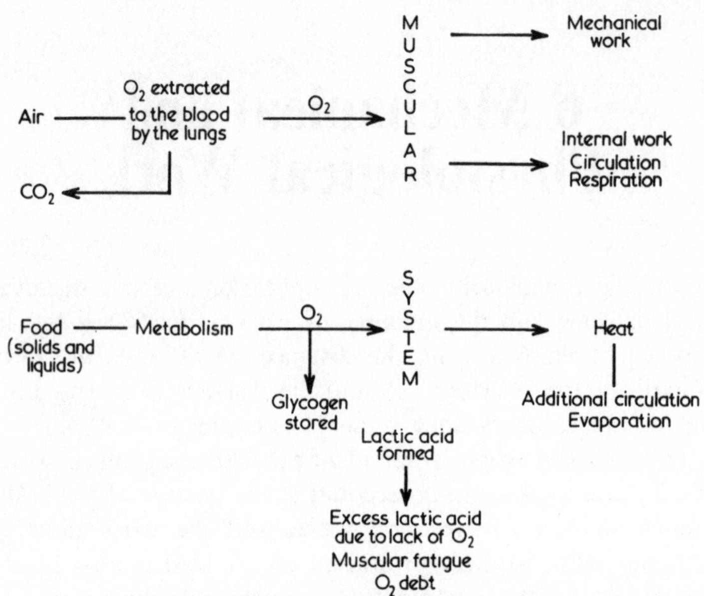

Fig. 10 Conversion of air, food and drink.

Fig. 10 gives a somewhat simplified explanation of the system's conversion of inputs of air, food and drink. The processes involved are highly complex and I shall not attempt to discuss them in any detail. The main energy conversions are:

- *Lungs*: Air is breathed; oxygen is transferred to the pulmonary blood; and evaporative cooling takes place.
- *Muscles*: Oxygen in the blood is changed into CO_2, with the formation of lactic acid when the oxygen supply is inadequate. Lactic acid produces muscular fatigue.
- *Body surface*: At Dry Bulb ambient temperatures up to 27°C (81°F) and with moderate exercise heat is lost from the body, 75% by convection and radiation and 25% by evaporative cooling by the lungs. With heavy exercise and above 27°C, evaporative cooling by perspiration is responsible for the main heat loss. The efficiency of this is proportional to the area of skin exposed and the air movement and is inversely proportional to the relative humidity.

- *Digestive processes*: Food and drink are absorbed by the system at a steady rate; the food provides a storage system and the drink maintains the water balance.

Units in physiological work

The energy expended, physiological work or physiological cost is related closely to the oxygen consumed. We can measure this directly in litres / minute or indirectly in heartbeats / minute. The basic unit used is the calorie expenditure in gram calories / minute.

Also Åstrand and Christensen investigated energy expenditure and the rate of the heartbeat and found that there was a direct relationship between them*. We can take advantage of this and use pulserates or heartbeats per minute to measure energy expenditure.

Energy levels

We are interested in three general levels of physiological work: resting, limit of aerobic work, and anaerobic work.

The resting level is the energy expenditure required to keep the body alive and is known as the Basal Metabolic Rate. It is measured by comparing the oxygen intake to the lungs and CO_2 output. The subject is starved for 12 hours before taking the test and is in a reclining position. Body weight and surface area are significant factors and the rate is normally stated in kilo calories / surface area / hour. For the average man weighing 65kg and having a surface area of 1.77m^2 it can be taken as 1 kilo calory / minute.

Work is said to be aerobic while the oxygen supply to the muscles is sufficient. Once this supply including reserves is inadequate, the system runs into 'oxygen debt': the work is

*Åstrand, P. O., *Experimental Studies of Physical Working Capacity in Relation to Sex and Age,* Ejnar Munksgaard, Copenhagen (1952); Christensen, E. H., 'Physiological Valuation of Work in Nykroppa Iron Works,' *Symposium on Fatigue*, H. K. Lewis, London (1953).

then anaerobic. Typical activities and their levels are shown in Table VII.

There is of course a physiological limit for any activity. This depends on skill, strength and state of health and can be increased with training. There are many examples in sport and athletics of the continual breaking of records and the success which scientific selection and training produce.

Mechanical and Physiological Work

It is not a practicable proposition to measure the mechanical work being performed by a man in industry. In any case, it would be of little value, as the physiological cost is what is required and this is not equal to but greater than the mechanical work due to postural work, environmental factors and so on. What we can do is to set up in the laboratory experiments on work loading under controlled conditions. Equipment has been developed for the purpose; the principal items involving mechanical work are the bicycle ergometer and the treadmill.

The bicycle ergometer is based on the static bicycle found in the gymnasium with instruments added to measure the mechanical work being done in kilogram meters per minute.

The treadmill is a modern development of that used in Victorian times in British prisons. (An example of a Victorian treadmill can be seen at St Helier, Jersey, Channel Islands. The diameter of the drum on which the prisoners trod was large enough to provide a practicable surface.) The modern laboratory treadmill resembles a belt drive, with the upper surface of the belt providing the tread upon which the subject walks, runs or bicycles. It enables a person walking or running to do so and remain stationary. Motor driven within a range of measured speeds, it can also be set at slight inclines to load the subject additionally. In case of a fall or collapse the subject will be thrown off the treadmill backwards and the safety devices to prevent this have to be carefully designed.

TABLE VII

Activities and Energy Levels

Energy (kcal/min)	1	2.5	5	7.5	10
Heartbeats/min	60	75	100	125	150
Oxygen (litres/min)	0.2	0.5	1	1.5	2
	Basal metabolism Resting Sleep Reclining	Light work Seated Driving an automobile	Walking at 6.5 kph Pushing wheelbarrow with 100kg	Hard housework Coalmining	Tree felling Furnace tending Moonwalk by Armstrong July 1969 (pulse 160)

Aerobic ←—|—→ Anaerobic

Basis: Standard man, weight 65kg, body surface 1.77m^2, energy reserve 25kcal.

Both the bicycle ergometer and the treadmill feature prominently in Brouha's discussion of physiology in industry*.

A third useful piece of equipment is the force platform. This is a triangular platform on which the subject is placed. At the three corners pressures are continuously sensed in the vertical, frontal and transversal directions so that reactions to all movements which he makes can be recorded in kilograms force by a three-pen recorder.

This is a useful apparatus when the work situation can be contained in this small space and for the comparison of different motion patterns in such tasks as bricklaying and stacking stores at awkward heights and in awkward postures. Certain areas of athletics, such as the golf swing and other activities where the athlete remains stationary, can also be studied.

At the same time as force units are being recorded physiological measurements are taken, and by analysis the work can be designed to optimize physiological cost.

*Brouha, L., *Physiology in Industry*, Pergamon Press (1960).

7 The Physiological Work Cycle

Heart Rate and the Physiological Work Cycle

If the heart rate is monitored during a cycle of rest, work and recovery, the period of recovery to the resting rate becomes increasingly longer with the intensity of the physiological work (fig. 11). In an extreme case the worker may not have fully recovered when he has to report for the next shift and will

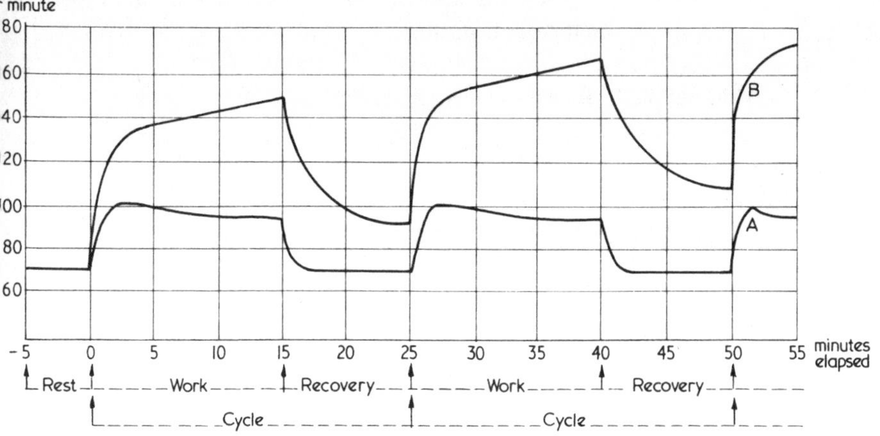

Fig. 11 The graph shows the difference in the body's reaction to light work and heavy work: during light work (curve A) the heart rate remains steady and recovers quickly during periods of rest; during heavy work (curve B) the heart rate rises sharply and does not recover the comfortable rate of curve A during rest. (Adapted from Brouha, L., *Physiology in Industry*, Pergamon Press, 1960.)

suffer from chronic fatigue. This is due to the fact that his oxygen debt was not repaid.

Murrell* suggests a method of optimizing the work and rest periods of the physiological work cycle. Assume that:

Energy reserve = 25 kcals
Average energy expenditure during recovery = 1.5 kcals/min
Standard energy cost, S, above which work becomes anaerobic = 5 kcals/min.

Let

w = total minutes in the shift,
a = time to recover in minutes/shift,
C_1 = work period in minutes,
C_2 = rest period in minutes and
b = Steady-state energy cost in kcals/min averaged including build-up.

*Murrell, K. F. H., *Ergonomics* (pp374–378).

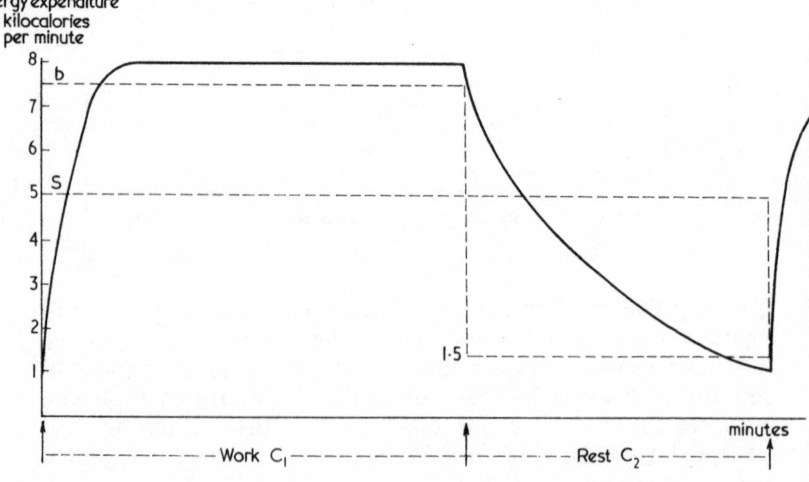

Fig. 12 Graph showing a physiological work cycle.

We can see from fig. 12 that for work periods the reserve of 25 kcals is used up according to

$$(b - S)C_1 = 25;$$

$$\text{i.e., } C_1 = \frac{25}{(b - S)} \qquad \text{. . . (i)}$$

For rest periods the reserve is replaced according to

$$(S - 1.5)C_2 = 25;$$

$$\text{i.e., } C_2 = \frac{25}{(S - 1.5)} \qquad \text{. . . (ii)}$$

But we require the total rest time or time to recover in minutes/shift. Now,

a = no. of rest periods in the shift × length of each rest period;

$$\text{i.e., } a = \frac{w}{C_1 + C_2} C_2 \qquad \text{. . . (iii)}$$

Adding (i) and (ii) we have

$$C_1 + C_2 = \frac{25}{b - S} + \frac{25}{S - 1.5},$$

$$= \frac{25(S - 1.5) + 25(b - S)}{(b - S)(S - 1.5)}$$

$$= \frac{25S - 1.5 \times 25 + 25b - 25S}{(b - S)(S - 1.5)}$$

$$= \frac{25(b - 1.5)}{(b - S)(S - 1.5)} \qquad \text{. . . (iv)}$$

Substituting (iv) and (ii) into (iii) gives us

$$a = \frac{w(b-S)(S-1.5)}{25(b-1.5)} \times \frac{25}{S-1.5}$$

$$= \frac{w(b-S)}{b-1.5} \text{ minutes.} \qquad \ldots \text{(v)}$$

Example

A task of average cost 7.5kcals/min is organized in periods of 10min work followed by 5min rest.

Number of periods/shift = 32.
Total shift = 8 hours.

What would you recommend as an ergonomic rest time? Assume the standard above which work becomes anaerobic is 5kcals/min.

$$C_1 = 10,$$

$$C_2 = \frac{25}{S-1.5} = \frac{25}{5-1.5} = 7.14.$$

Thus number of periods in the shift $= \frac{8 \times 60}{17.14} = 28$ recommended and rest time in the shift $= 28 \times 7.14 = 200$ minutes recommended;

$$\text{i.e., } \frac{200}{480} = 0.42 = 42\%.$$

Using Morrell's formula (v) directly we have

$$a = \frac{w(b-S)}{b-1.5} = \frac{8 \times 60(2.5)}{6} = 200 \text{ minutes.}$$

The Recovery Curve etc.

To avoid the disadvantages of measuring a man while working we can use the pulse rate changes during recovery. This technique was developed by Johnson, Brouha and Darling at the Harvard Fatigue Laboratory.

After a period of work the subject sits quietly and his pulse is taken at intervals from ½ to 1 minute after ceasing work, and similarly from 1½ to 2 minutes and from 2½ to 3 minutes. The heart rate recovery curve is then drawn and indicates:

- □ the physiological stress;
- □ the physical aptitude of the subject;
- □ the presence or absence of physiological fatigue; and
- □ physiological fatigue when a series of work periods are studied.

Examples of pulse rates using this technique are given in fig. 13 and are interpreted according to these rules:

- □ *For normal recovery*: Fall from first rate (*a*) to third rate (*c*) equals or exceeds 10 beats/minute. All three pulse rates, *a*, *b* and *c*, are equal or less than 90 beats/minute.
- □ *No recovery*: Fall from first to third rate equals or is less than 10 beats/minute. The third pulse rate is above 90 beats/minute.

Accordingly, falls of 10 beats/minute and third pulse rates at 90 beats/minute indicate critical points between satisfactory and unsatisfactory conditions.

Physiological Work and Rest Periods

As we have seen, body motions can be measured in mechanical work units, kilogram meters. However, the content and design of the task, the maintenance of balance, the braking of a load in motion, and awkward posture determine the physiological stress in reality.

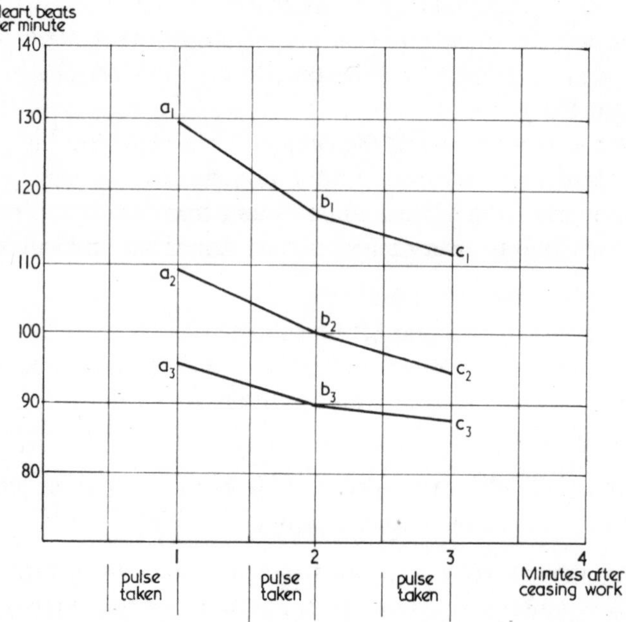

Fig. 13 Measurement by recovery curves (adapted from Brouha, L., see footnote, p60).

Despite mechanization a lot of manual lifting and manipulation of loads is required in all countries. This has been studied by the International Labour Conference, Geneva, and a report was published in 1964*.

Especially undesirable are:

□ lifting above eye-level;
□ pulling and pushing at low levels;
□ the cantilever effect; and
□ grasping badly-shaped loads.

*'Maximum Permissible Weight to be Carried by one Worker', *Occupational Safety and Health Series No. 5*, ILO.

Examples are stacking at height, opening and shutting the bottom drawers of a filing cabinet and skimming the surface of a tank with a long ladle at shoulder height. This last is an example of the cantilever effect.

Fig. 14 illustrates a case where there was a combination of the cantilever effect and lifting above eye-level. The worker had to handle work in and out of an electroplating tank; the work was suspended from rails running the length of the tank. The wrist joint was above eye level and he had to stretch up to clear the tank wall. Lifting and lowering vertically and keeping the load from striking the rails forced him to adopt a bending posture. The cumulative effect of repeating this task many times a day produced injuries to the spine and lower back.

The problem of making recommendations about maximum permissible weights is very complex and the whole matter is

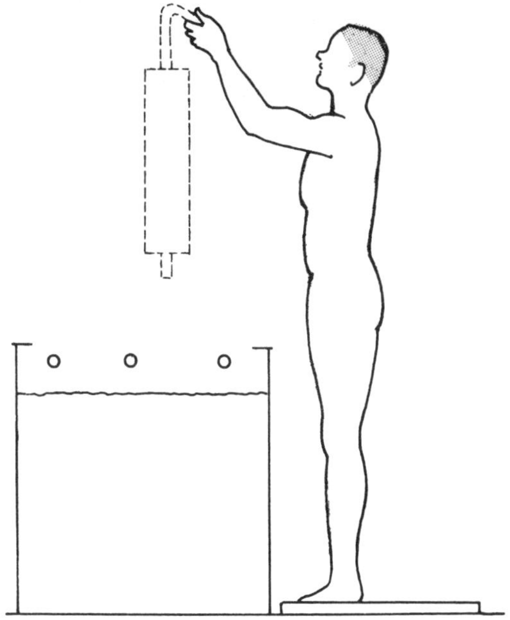

Fig. 14 A badly designed work position.

discussed by P. J. Carter in *Work Study and Management Services*, pp362–365, London (June 1969). All workers who have to lift and carry weights ought to be trained in the principles of correct lifting, which are shown in fig. 18 (page 90).

Considering physiological fatigue, we must organize the rest periods so that the workers' reactions at the end of the work shift are as similar as possible to those at the beginning. Short work periods followed by short rest periods can often prevent the accumulation of oxygen debt, as already discussed. A change of environment is also desirable, especially to a cool room in a hot environment and *vice versa.* This is, moreover, phychologically valuable.

8 Climate and Comfort

If the problem is to keep cool or warm, man is comfortable providing that:

$$M \pm R \pm C_1 \pm C_2 - E = 0,$$

where

- M = heat generated by metabolism;
- R = radiant heat;
- C_1 = heat transferred by convection;
- C_2 = heat transferred by conduction; and
- E = heat lost by evaporation.

As comfort is a subjective sensation, it is difficult to measure. Most people can tolerate an excessive dry heat, but find hot humid weather intolerable. However, people differ. Also the countries they live in and their climates influence their tolerance to heat and cold.

Heat Loss by Sweating

Our main defence against heat is sweating, E in the equation above. Evaporation of sweat causes an increase in blood flow to the skin and in the activity of the sweat glands. Evaporation takes place in the lungs and from the body surface, but the effectiveness of the system depends upon an adequate volume of dry cool air reaching the lungs and skin.

Cooling by air movement is less effective as the humidity and air temperature increase. If the air is above normal body temperature (37°C (98°F)) its movement ceases to be effective in removing heat from the body.

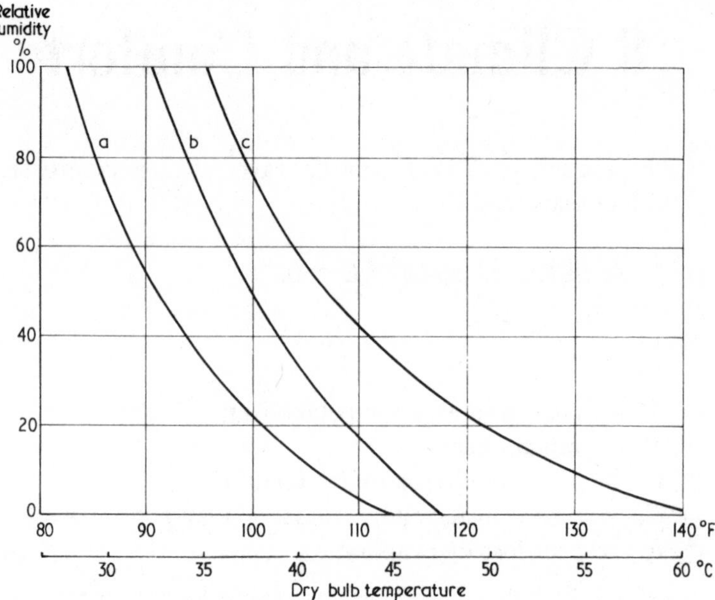

Fig. 15 The graph shows the upper limits for sweating: *a* is working nude; *b* is resting nude; *c* is resting nude with air movement. Points falling to the left of the curves represent tolerable conditions; points falling to the right represent intolerable conditions; clothing moves the curves to the left. (See Winslow, C. E. A., and Herrington, I. P., *Temperature and Human Life*, Princeton University Press, Princeton, NJ, 1949.)

The effects of heat loss by sweating are shown in Fig. 15*. In interpreting this, allowance has to be made for any clothing worn which will move the curves to the left. Working above the limits will cause the body temperature to rise beyond the normal and when it rises above 39°C (102°F), there will be a loss in working efficiency. Brain damage occurs above 41°C (106°F) and temperatures above 43°C (109°F) can be lethal after a few minutes. These temperatures are rectal tempera-

*Winslow, C. E. A., and Herrington, L. P., *Temperature and Human Life*, Princeton University Press, Princeton, NJ (1949).

tures, which are a consistent measure of deep body temperature.

The body's defence against cold is manifested as gooseflesh followed by shivering. Gooseflesh, the erection of hairs on the skin, is brought about by a narrowing of the arteries. It is more effective in the animal and bird worlds, where the hair standing on end and the ruffled feathers really do move the cold layers of air further from the skin. The next stage, shivering, activates the muscles to produce heat; in the days of the horse cab, the cabbies used to flail their chests with their arms to keep warm —much more effective.

Due to the narrowing of the arteries to make gooseflesh, the flow of blood to the extremities also is reduced. Thus the critical factors are often the maintenance of manual dexterity and ability to feel the feet. It is necessary to keep the fingers above 16°C (60°F) skin temperature to continue working without difficulty. Actual survival becomes critical below 28°C (82°F) internal body temperature*.

Thermal Comfort

We have to consider thermal comfort from two aspects. On the one hand, there is the objective aspect, the environment which can be evaluated and measured in terms of dry bulb temperature, humidity, air movement and radiant heat. The other aspect is subjective, the subject's sensation of heat or cold. As defined by ASHRAE for their New Effective Temperature Scale, this is 'that state of mind which expresses satisfaction with the thermal environment'.

*American Society of Heating, Refrigerating and Air-Conditioning Engineers (ASHRAE), *1977 Fundamentals Volume*, Chapter 8: Physiological Principles, Comfort and Health, ASHRAE Handbook and Product Directory, New York (1977). The referred chapter is an up-to-date and comprehensive account of important research in the factors relating to thermal comfort. The New Effective Temperature Scale (ET), its basis and applications are explained.

For the environmental factors, humidity or the water vapour content of the air can be measured with a sling psychrometer. This consists of two standard thermometers mounted so that they can be whirled through the air. One of the thermometer bulbs is dry and the other is covered with wet muslin. The instrument is slung around until the temperatures shown on the two thermometers are steady. The wet bulb temperature will fall depending upon the rate of evaporation of water from the muslin surface. The difference between the two is then noted. From a psychrometric table or chart various measures relating to moisture in the air can then be determined. Relative humidity (rh) is the measure generally used: this is the ratio of the amount of water vapour actually in the air to the maximum amount which can be held at saturation point at a given temperature and pressure. The saturation point is that point at which water vapour condenses as dew when the air is cooled.

Air movement is measured with a rotating vane device or by the cooling effect on a hot wire or on a special type of thermometer.

Radiant heat is discussed in a subsequent section.

The object of the ASHRAE scale and the many others which have been devised is to relate the two aspects of thermal comfort, the subjective and the objective, sensations of temperature and the environmental factors responsible.

The New Effective Temperature Scale

Research for the new ET scale was done at the Kansas State University in the USA. Eight hundred female and eight hundred male students, between 18 and 24 years of age, were exposed to dry bulb temperatures ranging from 15.6°C to 36.7°C and at relative humidities from 15% to 85%. The mean radiant temperature was equal to the dry bulb temperature and the air movement was less than 0.17m/s. All subjects were lightly clothed; they were allowed to read, play cards or

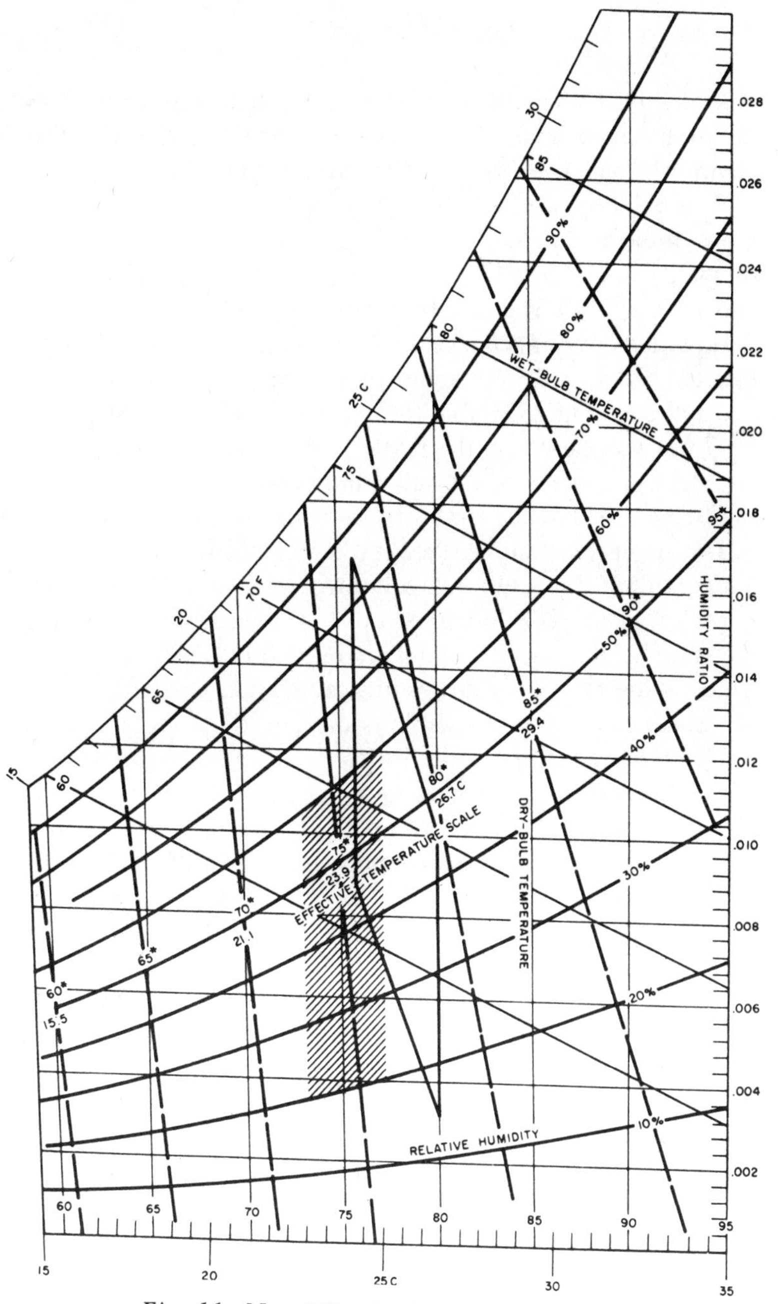

Fig. 16 New Effective Temperature Scale.

talk quietly. During three-hour sessions, each was required to say at specified intervals how hot or cold he or she felt. The choice was one of seven categories from cold (−3) to hot (+3).

Results showed that after three hours' exposure, one or more subjects felt comfortable over dry bulb temperatures ranging from 16.7°C (62°F) to 36.6°C (98°F): the average was 26°C (79°F). Relative humidity was 50%.

Equations were also developed to predict the thermal sensation to be expected from a given ET. The standard established and published in 1974 was the ASHRAE Comfort Standard 55–74; it was based on the research described, other research and field judgments of the subcommittee members.

The new ASHRAE comfort chart is shown at Fig. 16.

The environmental parameters in the chart are dry bulb temperature, wet bulb temperature, relative humidity and humidity ratio. The humidity ratio is a measure of specific moisture in the air; it is another term for specific humidity which is the mass of water vapour in relation to a given mass of air. It is usually expressed in grams per kilogram (g/kg).

The ET scale runs along the 50% rh line and the broken lines (----) crossing the rh lines show points of equal or constant physiological stress. For example, 31.7°C (89°F) dry bulb temperature at 20% rh feels as uncomfortable as 28°C (82.4°F) at 70% rh and the ET numerical value is 29.4°C (85°F).

Two comfort zones are drawn on the chart, a shaded zone and a diamond shaped envelope. The ASHRAE Comfort Standard 55–74, the shaded zone, applies to homes, schools, shops, theatres, etc. with average clothing. The envelope applies to lightly clothed workers (in shirtsleeves) seated at desks. The optimum in practice lies where the two zones overlap:

- ☐ ET 24.5°C (76°F);
- ☐ dry bulb temperature 24.5°C (76°F);
- ☐ 40% rh; and
- ☐ air velocity less than 0.23m/s.

Following the chart from cold to hot, ET can be taken as uncomfortably cold left of 15.5°C (60°F), the neutral point as 25°C (77°F) and beginning to be embarrassingly hot right of 35°C (95°F).

Radiant Heat

The hot humid conditions typical of laundries and ships' galleys and engine rooms in tropical seas can be improved by air conditioning. Air conditioning has, however, little if any effect in mitigating radiant heat. Sources of radiant heat are common in industry: furnaces, foundries, glass making and certain heavy welding processes.

The globe thermometer is used for measuring radiant heat; this consists of a hollow metal sphere of approximately 12cm diameter with a matt black outside surface and a glass thermometer with the mercury bulb at the centre of the sphere. It is placed at the point where the worker suffers most from the radiant sources, and its reading when stable, which will be well above the ambient air temperature, enables the mean radiant temperature to be calculated taking into account also air movement and dry bulb air temperature*.

The most simple and practical means of protection are shields of aluminium foil with windows of suitably shaded glass. Goggles and clothing are also used but interfere with the body's cooling unless very carefully designed or provided with forced ventilation.

*Bedford, T., *Environmental Warmth and its Measurement*, MRC War Memo No. 17, HMSO, London (1940).

9 Clothing

The discussion of clothing can be complicated by considerations of culture and fashion. I shall be confining my remarks to the purely ergonomic aspects.

With a few exceptions the traditional working dress of most countries has evolved according to its suitability to the local conditions: the large straw hat that wards off the radiant heat of the southern sun and the white loose clothing that reflects the Sun's rays and provides ventilation are typical examples. In the Amazon jungle, with its humid heat and lack of direct sunlight, as little clothing as possible provides the best conditions for evaporative cooling; hence the minimal covering worn there by the natives.

Insulation

Heat conduction through clothing is proportional to the temperature difference between the skin and the ambient air and inversely proportional to the volume of air trapped in the fabric.

These two relationships are especially significant in a cold environment and the resulting clothing will be bulky and likely to cause problems at the man / machine interface. Designers of seating, for example, have to give especial consideration to providing additional seat area and supporting configuration for the back and sides of the thighs. There will be greatly reduced feel both in the seating and also reduced grasp and feel in control levers, wheels, etc. The various control devices need to be designed with these problems in mind.

To combat wind chill, wind- and waterproof material is required; this at the same time may produce problems of dampness building up in the fabric between the skin and outer layer. In extremely cold climates the inner layers of clothing can become frozen during rest periods. What the clothing designer requires is a wind- and waterproof fabric which will also pass vapour.

Hot Environments

Neglecting for the moment radiant heat, we see that the effect of the two relationships of the previous paragraph are the opposite in a hot environment. Generally skin surfaces will be lower in temperature than the ambient air and the problem is to remove metabolic heat from the body. The surface-to-volume ratio is highest in the arms and legs, which therefore ought to be uncovered.

As evaporative cooling by perspiration takes place from all parts of the body, what minimal clothing as is worn must be permeable by water vapour. Vapour permeability is governed both by the type of fibre used and by the thickness of the weave.

Fibres rank in decreasing permeability as follows:

- Cotton
- Wool
- Nylon
- Rayon
- Glass

The basic hot-weather day uniform of the British Army at the end of World War II for general (not combat) use consisted of thick flannel wide-brimmed hats or topees; bush shirts of cellular cotton with short sleeves; wide, loosely cut shorts; wool-cotton mix stockings; and boots, shoes or sandals. This very successful uniform resulted from ergonomic studies.

Direct sunshine

The ultraviolet (UV) rays of the Sun are greatly respected in hot countries. However, they have low powers of penetration and a thin cloth will provide protection. On the other hand, the eyes are susceptible to strain from UV and too much exposure of the skin will produce burns. Production of Vitamin D is stimulated by UV taken in moderation; too much will destroy Vitamin D. Clothing ought to protect the head and back of the neck from direct sunshine.

The serious condition which used to be described as sunstroke and attributed to the effect of direct sunlight is now known to be caused by disorganization, and in extreme cases failure, of the body's heat regulating system. This failure can result from any condition of work and environment with which the normal cooling by evaporation of sweat cannot cope. The danger sign in cases of heatstroke, as it is today more correctly described, is when sweating stops and the body temperature rises to between 40.5 °C (105 °F) and 42 °C (108 °F). Immediate transfer to a cool environment is essential.

Cold Environments

The growth of the refrigeration industries and the arrival of the oil drills in Alaska make the cold environment of increasing interest to ergonomists. The chief hazards are frostbite and the freezing of exposed flesh. In calm conditions with still air, low temperatures are tolerable, but the combination of low temperature and wind is uncomfortable and, if severe, dangerous.

The National Weather Service of the USA has established an indicator to assess the relative discomfort caused by combinations of wind and temperature. The *wind chill index* or equivalent temperature is a measure of cooling power for various speeds ranging from 6km/h to 100km/h at temperatures from 20°C to −60°C (Table VIII). The value of 6km/h is selected as starting point for the table because a person walking briskly in calm air will generate a wind speed of about this.

One can see from the table that, for example, a wind speed

of 20km/h at −16°C has the same cooling power as the higher wind speed of 50km/h at a temperature of −8°C.

When the index has a low value it is customary in many of the states of the USA to broadcast warnings: one is well advised to take heed of these.

Dust, Fumes and Poisons

In industry, there are many conditions where discomfort, feelings of poor health of a temporary nature and also cumulative damage to health are caused by dust, fumes and poisons. In the case of processes which produce dust and fumes, separate enclosures with exhaust systems, or exhaust systems with mouths that suck away effluvia close to their sources, can be used. Also, workers can wear masks which filter out the undesired particles in the atmosphere or, in severe cases, masks which are arranged to provide entirely separate clean air to breathe.

An example of this in my own experience was the problem of sandblasting armoured vehicles to remove paint and rust. The sandblast guns had to be hand-held and directed at the surfaces and into interstices with reasonable accuracy to be effective. We found a satisfactory solution by providing the men with separately ventilated suits.

Poisons can enter the body by being inhaled, swallowed and absorbed through the skin. Some poisons the body can handle without damage; others accumulate in the body systems with eventual injury to health. It is these latter which the factory doctor must watch out for by checking the health of the workers concerned at regular intervals. Workers on certain processes—spray painters, for instance—can be provided with milk to assist the body functions.

When poisons are breathed they reduce the oxygen able to be absorbed into the blood stream by the lungs, and the lungs and heart have to work harder with a consequent rise in the

TABLE VIII

The Wind Chill Equivalent Temperature

WIND SPEED (km/h)

DRY BULB TEMPERATURE (°C)

	6	10	20	30	40	50	60	70	80	90	100
20	20	18	16	14	13	13	12	12	12	12	12
16	16	14	11	9	7	7	6	6	5	5	5
12	12	9	5	3	1	0	– 0	– 1	– 1	– 1	– 1
8	8	5	0	– 3	– 5	– 6	– 7	– 7	– 8	– 8	– 8
4	4	0	– 5	– 8	– 11	– 12	– 13	– 14	– 14	– 14	– 14
0	0	– 4	– 10	– 14	– 17	– 18	– 19	– 20	– 21	– 21	– 21
– 4	– 4	– 8	– 15	– 20	– 23	– 25	– 26	– 27	– 27	– 27	– 27
– 8	– 8	– 13	– 21	– 25	– 29	– 31	– 32	– 33	– 34	– 34	– 34
– 12	– 12	– 17	– 26	– 31	– 35	– 37	– 39	– 40	– 40	– 40	– 40
– 16	– 16	– 22	– 31	– 37	– 41	– 43	– 45	– 46	– 47	– 47	– 47
– 20	– 20	– 26	– 36	– 43	– 47	– 49	– 51	– 52	– 53	– 53	– 53
– 24	– 24	– 31	– 42	– 48	– 53	– 56	– 58	– 59	– 60	– 60	– 60
– 28	– 28	– 35	– 47	– 54	– 59	– 62	– 64	– 65	– 66	– 66	– 66
– 32	– 32	– 40	– 52	– 60	– 65	– 68	– 70	– 72	– 73	– 73	– 73
– 36	– 36	– 44	– 57	– 65	– 71	– 74	– 77	– 78	– 79	– 79	– 79
– 40	– 40	– 49	– 63	– 71	– 77	– 80	– 83	– 85	– 86	– 86	– 86
– 44	– 44	– 53	– 68	– 77	– 83	– 87	– 89	– 91	– 92	– 92	– 92
– 48	– 48	– 58	– 73	– 82	– 89	– 93	– 96	– 98	– 99	– 99	– 99
– 52	– 52	– 62	– 78	– 88	– 95	– 99	– 102	– 104	– 105	– 105	– 105
– 56	– 56	– 67	– 84	– 94	– 101	– 105	– 109	– 111	– 112	– 112	– 112
– 60	– 60	– 71	– 89	– 99	– 107	– 112	– 115	– 117	– 118	– 118	– 118

breathing and pulse rates. This can and does happen even from smoking a cigarette. Measurements can be made in the laboratory, and some experiments with small animals exposed to chemical environments have been interestingly described by J. A. Zapp Jr.*†

*See Footnote, p60.

†For the wide range of poisons in industry, read Buffa, E. S., quoting Patty, F. A., *Modern Production Management*, second edition, Chapter 10, Wiley Toppan, New York (1961).

10 Monitoring

Man-monitoring techniques using electrodes on the skin are used mainly in medical research and diagnoses, but they can also be useful in studying man in his working environment. The techniques which I am going to describe briefly are

- ☐ electromyography (EMG),
- ☐ galvanic skin response (GSR),
- ☐ electrocardiography (ECG or EKG),
- ☐ electroencephalography (EEG) and
- ☐ electrooculography (EOG).

These techniques are discussed in detail by Fogel and Murrell (see Footnotes, pp16 and 19 respectively).

EMG

If we place electrodes close to a muscle that is being used, measure the voltage (up to 500μV), amplify it and record it on a time base, we obtain a signal which varies as the effort expended by that muscle. The trace is fairly complex, with a frequency spectrum from 10 to 500Hz. By integrating the equation of the trace a linear relationship is found between the tension developed in the muscle and the integral. The use of this technique in work study has been discussed by Faulkner*. It is particularly applicable to work situations where the level of physiological effort is too low for measurement by study of heart rate or oxygen consumption.

Care is needed in applying EMG as the electrodes may pick

*Faulkner, T. W., *Work Study and Management Services*, pp742–747 (November 1969).

up voltages due to psychological factors in addition to or instead of muscular tension.

GSR

Connecting electrodes to both hands, we obtain a voltage which is a sensitive indicator of cortical and higher level functions. What we are measuring is how the conductivity of the body changes when a stimulus excites the sweat secretion reflex. This is an involuntary reflex produced by sudden fear and mental or emotional stress. It can also indicate levels of relaxation and drowsiness or alertness giving a picture of fatigue and competence of a driver of a vehicle or other worker after given periods at work* Its best known use is in the lie detector where words that are emotionally loaded to the subject produce an involuntary reaction.

ECG

This will be familiar to many as the technique used in the diagnosis and medical study of heart conditions. The normal pattern shown in the trace has been mapped and variations from this pattern can be interpreted by the expert. The pattern is affected both by physical and by mental or emotional stress. Its main use in ergonomics is in conjunction with the GSR for lie detection.

EEG

Placing electrodes on the scalp we obtain an indication of the total nervous activity taking place in the brain. The spectrum of frequencies and amplitudes of the various wave forms is analysed by identifying components of certain frequency

*Stevens, S. S., *Handbook of Experimental Psychology*, Wiley, New York (1951).

ranges. These are associated with a subject in deep sleep, resting with eyes shut and in a condition of mental stress.

The component most useful in ergonomics is the α rhythm between 8 and 13Hz. As this varies from person to person, the α rhythm index is used which measures the ratio between the α rhythms present in an individual during activity to those during rest with the eyes closed.

Using this approach, work which imposes nervous loading with low physical loading can be investigated and evaluated. There are more and more of this type of tasks in modern industry and the importance of their evaluation is increasing.

An interesting study of typical tasks has been made by Hartemann, Manigault and Tarrière*. In addition to recording the individual EEGs, heart rates were taken. Little evidence of α rhythms was found during the working day, indicating continuous high nervous activity. At the same time there was an appreciable increase in heart rate from morning to evening without relationship to the muscular or physical requirements, which remained low.

A somewhat unusual application of EEG is made by those people who wish to experience a 'high' without taking drugs. If the signal taken from the brain activity is amplified and fed back through ear plugs, the α rhythm is stimulated and maintained, giving sensations valued by some.

EOG

The study of eye movements and fixations on points in the working space aids the designer. Displays for pilots in aircraft and for control engineers on consoles in chemical and power plants and distribution networks can be checked for optimum layout in terms of frequency of fixation, distances moved and

*Hartemann, F., Manigault, P., and Tarrière, C., 'An Endeavour to Evaluate the Nervous Load at Work Stations in Line Production', *The International Journal of Production Research*, Vol 8, No 1, Institution of Production Engineers, London (1970).

relative importance of the information content. In the past this has been done by making a motion picture of the eyes or alternatively by attaching very small mirrors to the eyeballs which reflect beams of light. The more convenient method of electrooculography (EOG) is to use the electrical signals which electrodes attached above and below the eyes will give when the eye muscles are stimulated. A light mask accurately positions the electrodes on the face. The subject under study is little affected by the apparatus.

11 Posture, Lifting and Handling

Posture at work and related seating, tables and equipment have been studied in depth and at great length and the findings and recommendations have been widely published. Nevertheless, even today there are in industry many examples of job designs which produce excessive fatigue and often ill-health. In the same way, much ill-health and many accidents result from lifting, carrying and handling. Although we have dealt briefly with this subject already in Chapter 7, its importance merits returning to it once again.

Seat Design

Seat design aims to provide support so that the sitter can make use of a number of satisfactory postures. The following are to be avoided:

- ☐ restriction of blood flow;
- ☐ tension in muscle groups beyond the normal;
- ☐ undue pressure on the buttocks;
- ☐ convex curvature of the spine;
- ☐ too much control; and
- ☐ impermeable material as seat covering in hot humid climates.

Tables of desirable dimensions can be found in Woodson, and Murrell, Chapter 8†.

*Woodson, W. E., *Human Engineering Guide for Equipment Designers*, pp2–142 to 2–157, Berkeley University California Press (1954).

†See Footnote, p19.

Standing

Standing at work should be avoided. Where this is not possible, the position ought to be upright. Leaning over a bench is a common posture in industry; it involves a relatively high static work rate. The correct work height is level with the elbow, the arm hanging freely at the body side. The work-bench surface ought to be 8cm below this. Also 30cm clearance space is necessary for one foot to be placed in front of another to assist balance.

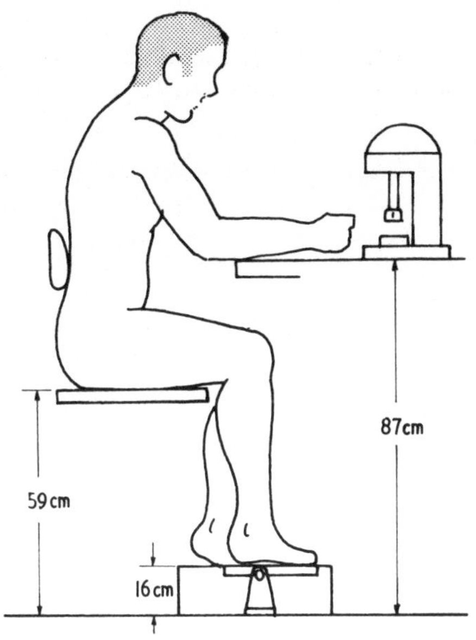

Fig. 17(a) Sitting: a good design. The worker is operating with a pedal control; a secure footrest must be provided for the other foot.

Examples of sitting and standing postures are shown in figs. 17(a) to (e).

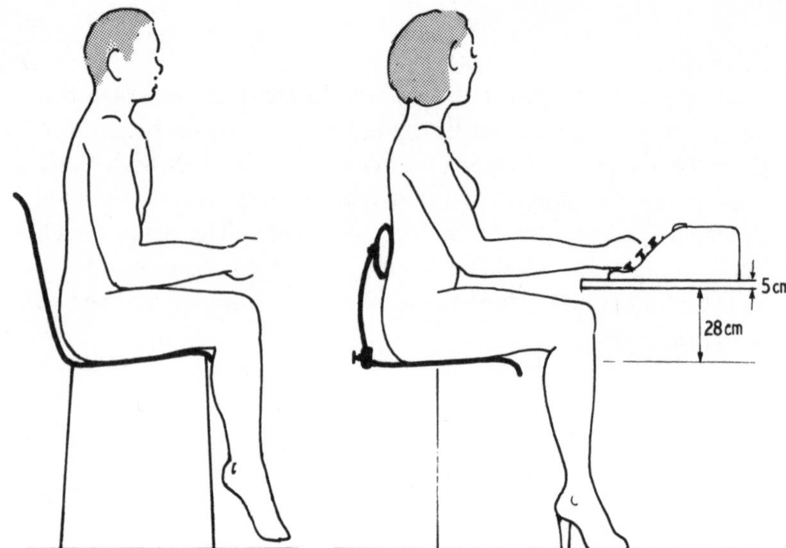

Fig. 17(b) Sitting: here the seat is too high, causing excess pressure on the buttocks, and too deep, so that the front of the seat cuts into the back of the knee restricting the blood flow. The lumbar region of the back is not adequately supported.
Fig. 17(c) Sitting: here the clearance between seat and desk so that the girl is not prevented from crossing her legs ought to be no less than 28cm; the thickness of the table should be 5cm. The centre of the back rest should be adjustable to somewhere between 18 and 25cm from the seat top to support the small of the girl's back.

Manual Lifting and Handling

Using incorrect lifting postures or attempting to lift too heavy weights can produce backaches, strained and pulled muscles and ruptures of the stomach wall. There are, however, established principles to be followed in lifting correctly. These are:

- correct grip, with palm of the hand and roots of the fingers and thumb;
- straight back, with the load close to the body, hips, knees and ankles flexed, and chin in to strengthen the back;

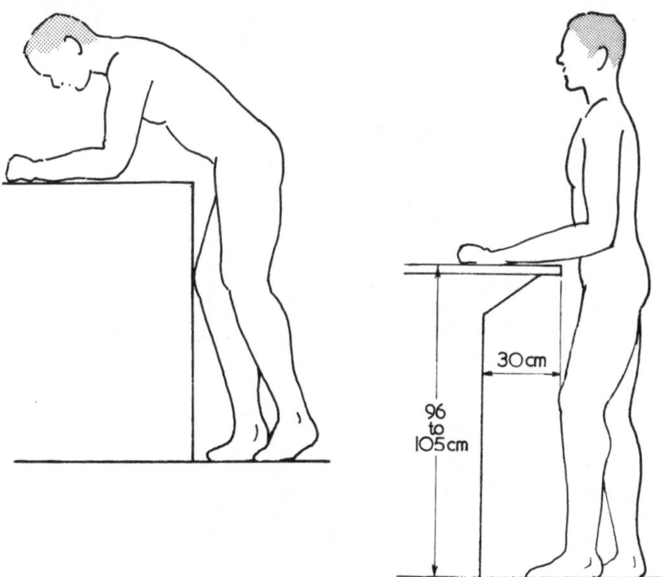

Fig. 17(d) Standing: the man is suffering from muscle fatigue and his spine is curved; there should be room under the desk for him to stand with one foot forward.
Fig. 17(e) Standing: a good upright position because there is room under the bench (at least 30cm). But there should also be some kind of stool, between 70 and 85cm high, and the bench top should not be more than 8cm beneath the level of the man's elbow.

- □ feet slightly apart, one ahead of the other; and
- □ arms close to the body.

In any situation where manual lifting is needed the following must be considered: frequency, weights and shapes to be lifted, distances, and opportunities for recovery.

The ILO has studied this question and issued its report*.

*See Footnote, p66.

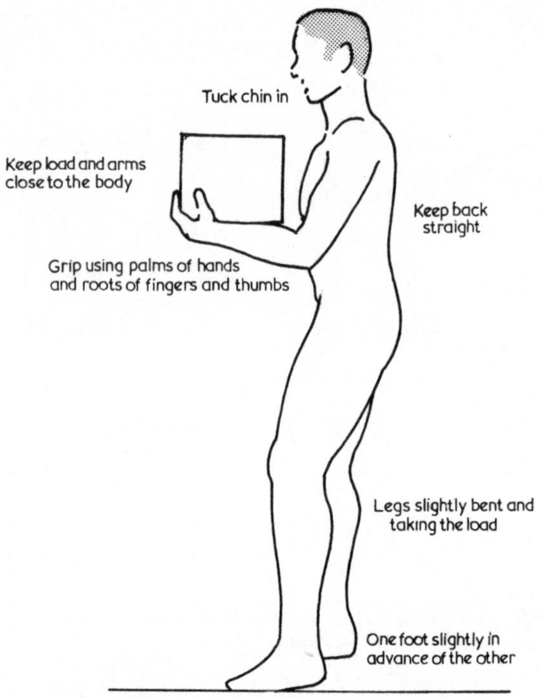
Tuck chin in
Keep load and arms
close to the body
Keep back
straight
Grip using palms of hands
and roots of fingers and thumbs
Legs slightly bent and
taking the load
One foot slightly in
advance of the other

Fig. 18 Lifting: a good posture for lifting and carrying includes keeping the back straight, the legs slightly bent to take the weight of the load, the arms—carrying the load—close to the body, one foot slightly in front of the other, and the chin in.

50kg was the maximum recommended weight for male workers. There were some 175 standards in numerous countries at that time and the matter is not as simple as it may seem*. A recommended posture for lifting and carrying is shown in fig. 18.

*More recent and helpful advice is contained in: *Lifting in Industry*, The Chartered Society of Physiotherapy, 14 Bedford Row, London (1974); and *Lifting and Carrying—Health and Safety at Work*, Booklet 1, Department of Employment, HMSO, London (1972).

12 The Home

'owhere has the impact of ergonomics been more felt in ecent years than in the home. Better architechural design, etter layout of equipment in the kitchen, increased safety and more convenient handles for anything from razors to mowing machines have been developed. Nevertheless, it is rare to find in the UK a house with windows whose exterior surfaces can be easily and safely cleaned from the inside by the housewife. Invariably windows are either difficult to clean or virtually impossible in practical terms for the unaided housewife, so specialist window cleaners have to be employed.

Another not uncommon defect is found in the design of doors and the positioning of door handles so close to the doorway into which the door fits that the hand is nipped on closing, trapped between the handle and the vertical member of the door case.

Sometimes good ergonomic design is not well received and does not sell. A toothbrush with its handle twisted through 90° so that the flat part of the handle was at right angles to the line of the bristles was placed on the market some years ago. This was more convenient to hold and ergonomically superior; however, it was not well received and had to be discontinued. Below are examples of ergonomic design and some of the factors which have to be considered.

Utensils

The design of many utensils used in industry and the home leaves much to be desired. Indeed, those for cooking and hot liquids can often be the cause of accidents.

The features of main ergonomic interest are:

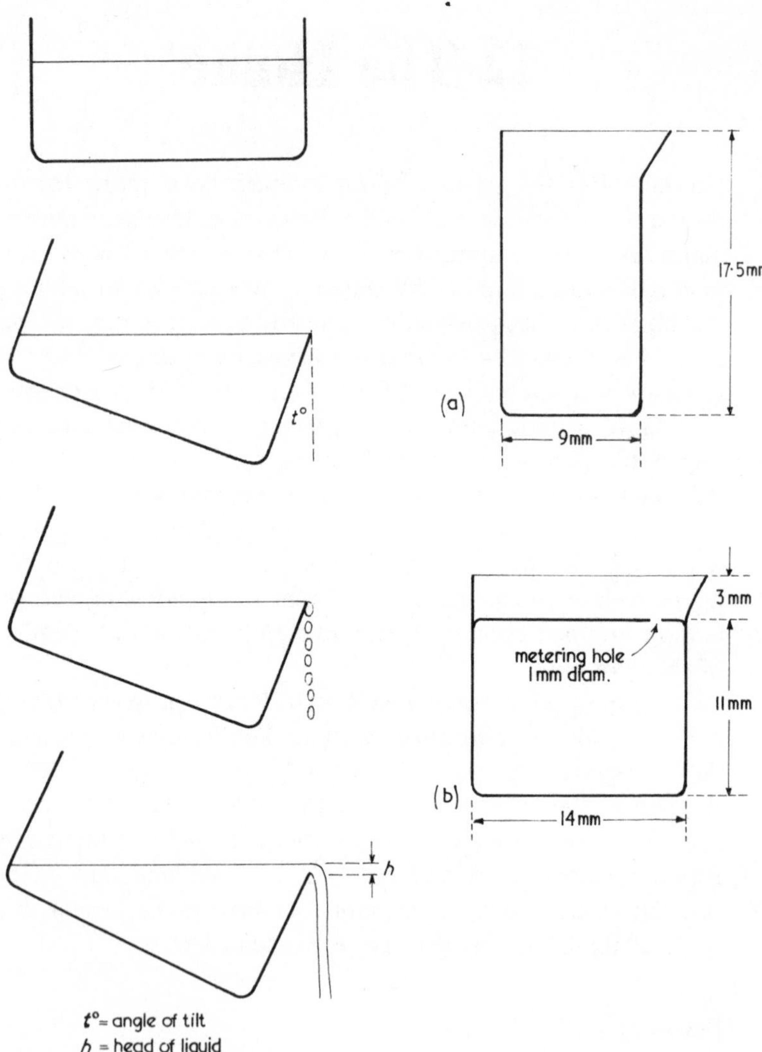

Fig. 19 Pouring action: $t°$ is the angle of the tilt and h is the head of the liquid.

Fig. 20 Utensils: (a) has a good spout which makes pouring easy; but is too narrow and deep for easy cleaning. (b) has a metering hole that fits 3mm from the top and makes pouring easy; the wide, shallow shape of the jug makes cleaning easy.

- □ the handle, convenient to grip, and heat-insulated when appropriate;
- □ the ease with which a liquid can be poured without spilling or dribbling; and
- □ the accessibility for cleaning of the interior.

The first feature can be readily assessed; some designs which use plastic handles on cooking pans, however, place the plastic material in such a position that it can be damaged by the heat source.

The second feature, which we can call 'pourability', is more difficult to judge from the appearance of the utensil.

Naturally, the viscosity of the fluid affects the ease with which it can be poured—water is easier to pour than porridge. Pouring starts as soon as the utensil is tilted into a position which provides a head (fig. 19), and the greater the volume per second the steadier will be the configuration of the jet. At low volumes per second, molecular attraction between the liquid and the material of the utensil is found to be an adverse factor. The angle of tilt can be increased as the utensil is itself tilted by suitable design (fig. 20); also a triangular notch is presented to the advancing fluid, shaping and controlling the shape and steadiness of the jet. The addition of a spout, as in the traditional British teapot (fig. 21a), or a metering hole (fig. 20b) gives even better control. The tricks which in Spanish-speaking countries the wine waiter does, directing a jet of wine into his mouth or onto the centre of his forehead, are made possible by a well-designed utensil with metering built into it. Frequently an attractive appearance acts against good ergonomics. Fig. 21b shows an example of an attractive utensil with extremely bad 'pourability'.

Accessibility is normally based on the ability to get at the principal interior surfaces with the hand; this is necessary for good hygiene but is not provided in some articles of crockery, in particular the smaller jugs. Fingers and a mop can, of course, be used, but not so effectively. The only sure way of

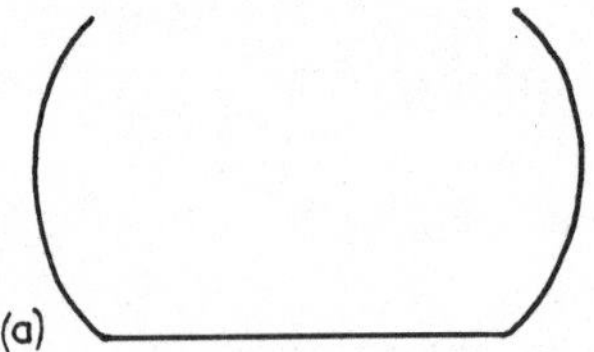

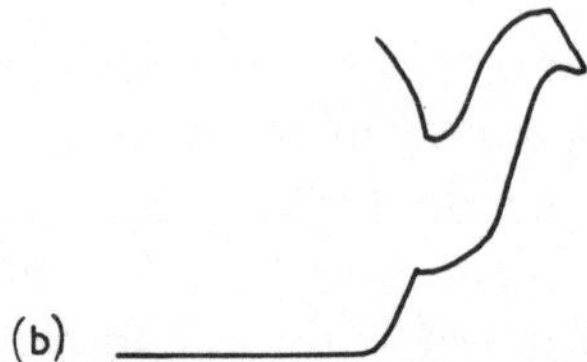

Fig. 21 Utensils: (a) is attractive but has poor 'pourability' whereas (b), with the added spout, has good 'pourability'.

buying satisfactory teapots, jugs and utensils which pour is to ask the sales assistant to fill it with water and demonstrate its ability to pour cleanly. However, you may make yourself unpopular with the sales assistant!

Handles

Let us take the meaning of handle in the general sense as ranging from anything which is grasped by the hand to that manipulated by the fingers, say from hammers to crochet hooks. Even confining our interest to the home, there is great variety in the numerous utensils and implements which are handled and variety also in the attitude adopted by the hands and configuration of the fingers and thumb. However, comfort and efficiency is readily evaluated by the user in each case. The general method of design is to shape the handle to conform in three dimensions to the pressure areas of the thumb, fingers and palm as they grip or manipulate the article.

Many tools have traditional shapes which are accepted as satisfactory, but in recent years better designed and more

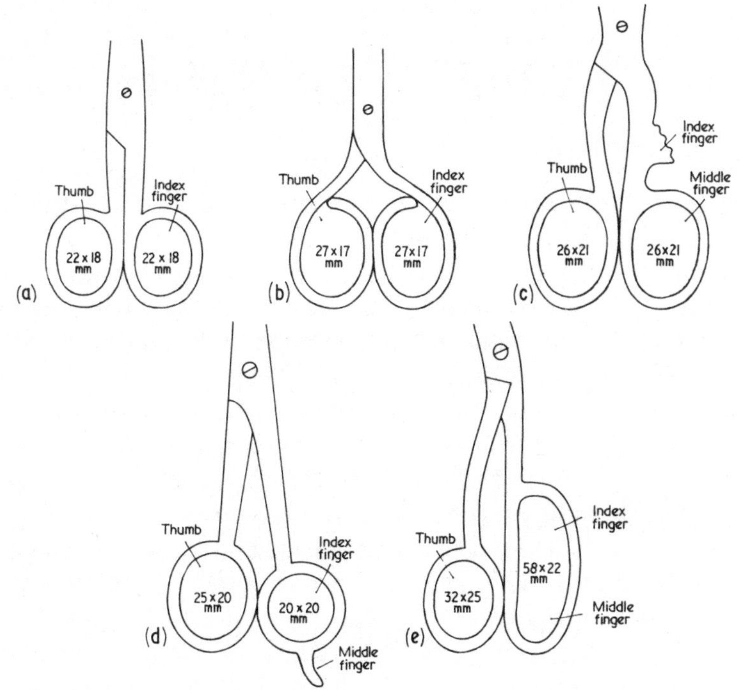

Fig. 22 Scissors: examples of thumb and finger grasps.

expensive articles have appeared in the shops and consumers have become more selective.

The scissors and shears shown in fig. 22 demonstrate different ways in which the design accommodates the thumb and fingers to give sensitive control and physical force, as usage requires. The traditional nail scissors (a) allow the thumb and index finger barely to enter, and provide sensitive control, while the scissors (b) cater for a stronger action. In (c), (d) and (e) both the index and middle fingers can exercise force. I have chosen these examples as they are present in most homes; anyone can explore and experience ergonomic design in action.

It is interesting that at long last a new design of thimble has been developed allowing the user to guide the needle positively and at the same time feel the texture of the material being worked.

13 Industrial Safety

Although my experience as a consultant in industrial safety has been concentrated mainly in the south of England, I have sampled a fair cross-section of industrial and other activities which provoke accidents. The variety of accidents seems infinite, and the equipment and environments associated with accidents also provide no common pattern.

Often poor ergonomic design of equipment is a contributory factor, but more often the layout and design of the place of work and physiological loadings can be faulted. Lifting and manipulation of loads by hand are still common today; in my experience, managements seem ready to accept job designs which place workers in difficult, even unsafe, postures, considering the weights involved and the frequency of lifting.

Despite designers' continuing efforts to design foolproof guards to protect workers from moving parts, cutters and dangerous operations, there is a high proportion of accidents from the guards being left off after repairs and manipulated or removed for the clearing of snags and cleaning of tools. The provision of fail-safe systems would prevent many accidents; effective are micro-switches which break the power circuit as long as the guard is not in position, and other similar devices. However, these have to be inspected and maintained rigorously.

The positioning of emergency stop switches is often unergonomic; they ought to be at the correct height and in an area convenient for the operator and at the same time near nips and possible hazards. Controls for inching and setting moving parts to setting points are often awkward and require the operator to perform physical miracles to enable him to see the work-point and move the control with the necessary precision at the same time. Management can often be criticized for buying machines with these faults: it can only be assumed that no user-trials are made before selecting such plant, or that better designs were not available.

The Health and Safety at Work Act (UK) came fully into force on 1 April 1975. Although it only partially replaces existing legislation, it provides for the repeal, amendment, revision and updating of existing acts, and regulations will be issued over the years to these ends. The scope of legislation is extended and the powers of inspectors increased. Duties and responsibilities which were formerly covered only by common law have become statutory. The end result of these developments is to focus on the actual work-place and the worker there; in fact, what ergonomics is all about.

Appendix

An Approximation with Sabine's Formula

A lecture theatre has unsatisfactory acoustic properties: the lecturer often hears his words repeated due to reflected sound. A plan and developed view of the walls are contained in Figure 23. (This case is based on actual experience.)

The following conditions apply:

Seating	150
Seated audience	100
Desired reverberation time	0.8 to 1.0 second

Absorbent material to be mounted above 1.8m
Neglect absorption at frequencies other than 500Hz
Assume absorption coefficients at 500Hz:

Walls	0.06
Ceiling	0.02
Floor	0.05
Wood surfaces	0.10
Window glazed	0.18
Absorbent material	0.70
Person seated	0.40 $S\alpha$ units per person
Unoccupied seat	0.16 $S\alpha$ units per seat

Method

- □ Estimate the absorption in Sabins present in the lecture theatre untreated;
- □ calculate the absorption required for the desired reverberation time;
- □ determine the total area of absorbent material required;
- □ plan the areas to be treated; and
- □ check the result.

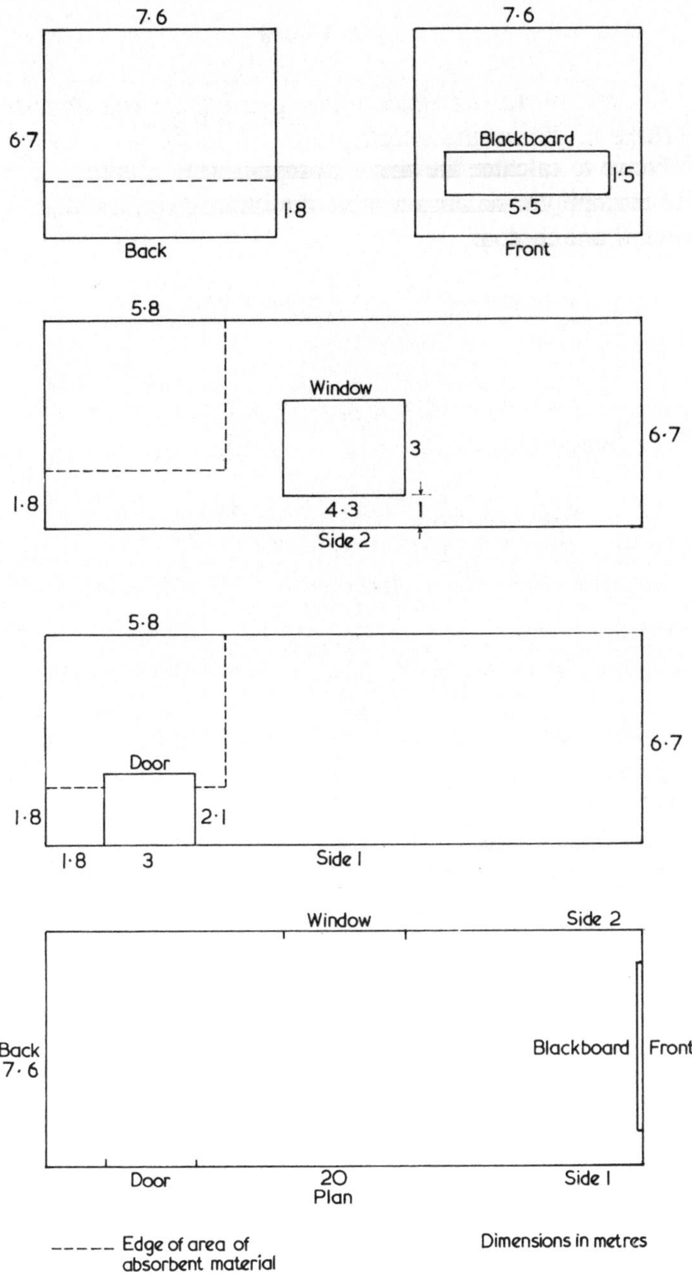

Fig. 23 The lecture theatre. (All measurements are in metres.)

1: Estimate the absorption present in the lecture theatre in its untreated state

We have to calculate the various absorption areas separately. For example, the surface area of all four walls, less blackboard, window and door, is:

$$\underbrace{2(7.6 \times 6.7)}_{\text{Ends}} + \underbrace{2(20 \times 6.7)}_{\text{Sides}} = 370\text{m}^2$$

$$\text{less } \underbrace{(5.5 \times 1.5)}_{\text{Blackboard}} + \underbrace{(4.3 \times 3)}_{\text{Window}} + \underbrace{(3 \times 2.1)}_{\text{Door}} = 27\text{m}^2.$$

$$\therefore \text{Wall surface} = 370 - 27 = 343\text{m}^2.$$

Surfaces	*Areas* (m^2)	*Coefficients*	*Sabins*
Walls	343	0.06	21
Ceiling	152	0.02	3
Floor	152	0.05	8
Blackboard	8	0.10	1
Door	6	0.10	1
Window	13	0.18	2
Seated Audience	100	0.40	40
Unoccupied Seats	50	0.16	8
			84

2: Calculate the absorption required for T = 0.9

But Sabine's formula,

$$T = \frac{0.16\,V}{\Sigma\, S\alpha} \text{ can be written } \Sigma\, S\alpha = \frac{0.16\,V}{T}$$

Substituting 0.9s for T and 1020m^3 for V we have $\dfrac{0.16 \times 1020}{0.9}$ = 181 Sabins, the total absorption required for a reverberation time of 0.9 seconds.

3: Determine area of absorbent material required

We already have 84 Sabins, therefore we need to arrange for additional absorption of $181 - 84 = 97$ Sabins.

$$\text{Area of absorbent material necessary} = \frac{97}{0.7} = 139\text{m}^2.$$

4: Plan the areas to be treated

If we cover the ceiling and walls above 1.8m with absorbent material to a distance of x metres from the back wall, then:

$$\underset{\text{Back wall}}{7.6(6.7 - 1.8)} + \underset{\text{Side walls}}{2(6.7 - 1.8)x} + \underset{\text{Ceiling}}{7.6x} = 139;$$

i.e., $37 + 9.8x + 7.6x = 139$,

$17.4x = 102$.

$\therefore\ x = 5.8\text{m}$.

For best results the material on the ceiling ought to be disposed along the edges; accordingly, cover the ceiling with three strips 1m wide, two 19.2m lengths extended along both sides from the back and the third a 5.6m length fill in at the back end of the ceiling.

5: Check the Result

With the addition of absorbent material, we now have:

Surfaces	*Areas*	*Coefficients*	*Sabins*
Absorbent			
Back wall	37.2	0.70	26.0
Ceiling	44.0	0.70	30.8
Side walls	55.9	0.70	40.0
Untreated			
Back wall	13.7	0.06	8.2

Surfaces	*Areas*	*Coefficients*	*Sabins*
Untreated			
Ceiling	108.0	0.02	2.0
Floor	108.0	0.05	5.4
Front wall	43.0	0.06	2.6
Side 1	99.4	0.06	5.9
Side 2	82.0	0.06	4.9
Blackboard and Door	14.0	0.10	1.4
Window	13.0	0.18	2.3
Seated Audience	100.0	0.40	40.0
Unoccupied seats	50.0	0.16	8.0
			177.5

$$T = \frac{0.16\,V}{\Sigma\, S\alpha} = \frac{0.16 \times 1020}{177.5} = 0.92 \text{ seconds.}$$

which is within the requirement.

Using coefficients of absorption for 125Hz and 2,000Hz, the reverberation times at these frequencies can be checked. At 2,000Hz, air absorbs sound and the appropriate absorption units ought to be added. If the room is found to be acoustically too dead in practice, some of the absorbent material can be removed*.

*For a discussion of the acoustic problems of rooms for speech, consult Anon, *Building Research Establishment Digest 192*, 'The Acoustics of Rooms for Speech', HMSO (1976). Also Smith, B. J., *Environmental Physics: Acoustics*, Longman Group, Harlow (1971), contains worked solutions of typical problems and covers the acoustics of environmental design.

Acknowledgements

Acknowledgement is due to the following for tables and figures: Table 1: adapted from Ralph M. Barnes, *Motion and Time Study* (1968), John Wiley and Sons Inc. Table 5: *Noise, Final Report Cmnd 2056* (1963), HMSO. Table 6: A. G. Aldersey-Williams, *Noise in Factories,* Factory Building Studies No. 6 (1960), HMSO. Table 7: National Climatic Center, Environmental Data Service, NOAA; and Meteorological Services Division, National Weather Service, NOAA, USA. Figures 7(a) and (b): R. L. Gregory, *Eye and Brain* (1966), Weidenfeld & Nicolson. Figure 7(b): L. S. Penrose and R. Penrose, 'Impossible objects, a special type of illusion' (1958), the *British Journal of Psychology* and Cambridge University Press. Extracts, p24, are from the *IES Code for Interior Lighting* (1977), published by the Illuminating Engineering Society (since incorporated into the Chartered Institution of Building Services). Figure 8: N. R. F. Maier, *Psychology in Industry* (1965), George G. Harrap & Co. Ltd. Figures 11 and 13: L. Brouha, *Physiology in Industry* (1960), Pergamon Press. Figure 12: Based on arguments put forward by K. F. H. Murrell, *Ergonomics, Man in His Working Environment* (1965), Chapman & Hall Ltd. Figure 15: C. E. A. Winslow and L. P. Herrington, *Temperature and Human Life* (1949), Princeton University Press; reprinted by permission of Princeton University Press. Figure 16: Reprinted, slightly modified, with permission from the *1977 Fundamentals Volume*, ASHRAE Handbook and Product Directory.

Recommended Reading

Blakemore, Colin: *Mechanics of the Mind*, Cambridge University Press (1977). The text of the BBC Reith Lectures of 1976, slightly expanded; a beautifully presented and written book, it makes fascinating reading. Application of EEG and interpretation of the results are studied.

Burns, William: *Noise and Man*, John Murray, London (1968). A comprehensive and authoritative text by the Professor of Physiology at Charing Cross Hospital Medical School, London. It deals thoroughly with psychological and organic effects and occupational hearing loss and its prevention. Modern noise pollution is also studied.

Carlsöö, Sven: *How Man Moves: Kinesiological Methods and Studies*, Heinemann, London (1972). An unusual book which gives an insight into the mechanics of the body. Examined is research work using EMG into various topics such as lifting, archery and the golf swing.

Chapanis, A., Cook, J. S., Lund, M. W. and Morgan, C. T.: *Human Engineering Guide to Equipment Design*, McGraw Hill, New York (1963). A comprehensive and authoritative text covering the matching of man and machine. Chapter 11, on human body measurements, is especially valuable.

Fitts, Paul M. and Posner, M. I.: *Human Performance*, Brooks-Cole, Belmont (Calif.) (1973). The many aspects of human performance are studied in this authoritative text. Motivation, reaction time, skill, memory, language and information processing are examples of subjects covered.

Gregory, R. L.: *Eye and Brain: The Psychology of Seeing*, Weidenfeld and Nicolson, London (1966). This attractive and beautifully illustrated book discusses in an interesting and readable manner how the eye and brain work together. The many ways in which the brain can be deceived are presented in a fascinating way.

McCormick, E. J.: *Human Factors Engineering*, McGraw Hill, New York (1957). A considerable text (637 pages) encompassing the field of human factors engineering in detail, and with clear line drawings and photographs. A similar book to Chapanis *et al*, above.

Mueller, Conrad G. and Mae, Rudolph: *Light and Vision*, Time-Life International (Nederland) NV, Amsterdam (1967). Written with a pleasing absence of scientific jargon and superbly illustrated, this book covers not only man but also the animals, ranging from historical to modern times as the topics of vision, light and colour are explored. Interesting and easy to read.

Murrell, K. F. H.: *Ergonomics: Man in His Working Environment*, Chapman and Hall, London (1965). Written by the English authority on ergonomics, this important and well organized book explores the full compass of ergonomics in an interesting and readable manner. Research is well covered and related to the sort of problems experienced in industry.

Renbourn, E. T.: *Physiology and Hygiene of Materials and Clothing*, Merrow, Newcastle-upon-Tyne (1971). This monograph explains how clothing works and describes interestingly the experimental work involved in developing and proving new designs and materials.

Sisson, C. H.: *Code of Practice for Reducing the Exposure of Employed Persons to Noise*, Her Majesty's Stationery Office, London (1972). Practical guidance to prevent loss of hearing at work. Much-needed in certain industries.

Tolansky, S.: *Optical Illusions*, Pergamon, Oxford (1964). A short, witty and entertaining discussion of geometrical optical illusions. Splendid illustrations.

Glossary

Afferent Nerves. Nerves that carry impulses to the central nervous system.

Atmospheric Perspective. Qualities of the atmosphere which suggest distance. Distant objects seen through clean air appear to be nearer than those seen through air containing smoke, dust, etc.

Audiometry. The measurement of hearing sensitivity. The subject listens to pure tones of various frequencies at known pressure levels and indicates which he can detect with each ear separately.

Autonomic System. The part of the nervous system which controls the involuntary action of the heart, the lungs and the processes necessary to maintain life in the body.

Axon. The main fibre which carries impulses from a nerve cell.

Cardiac Output. The total volume of blood expelled by the heart per minute. Clearly, the number of heart beats per minute and the volume of blood pumped with each beat determine the output.

Cantilever. A beam or girder fixed at one end and free at the other. A load on the free end is described sometimes as a cantilever load.

Cutaneous Receptors. Receptors in the skin for detecting touch, pain, heat and cold.

Dendrites. Small fibres extending from the body of the nerve cell. Impulses passing from one cell to a second go from the axon of the first through a synapse and then to the dendrites of the second cell.

Diurnal Rhythm. A 24-hour cycle. Many of the body's metabolic processes follow a regular 24-hour pattern.

Efferent Nerves. Nerves that carry impulses away from the central nervous system.

Ergometer. A device to measure physical work.

Exteroceptor. A sensory nerve-ending which receives impressions from outside the body.

Graphic Pen Recorder. An electrical device in which a pen (or several pens) moves over a paper chart so that a graphic record is obtained of a quantity or quantities measured.

Hertz (Hz). The unit of frequency, the number of times a cycle is repeated in a period of one second.

Inching. The slow and precisely controlled movement of a tool in a machine tool or other article.

Interoceptor. A sensory nerve-ending which receives impressions from within the body.

Masking. The loss of sensitivity of the ear to specified sounds in the presence of other sounds.

Metabolism. The sum total of all the chemical reactions that take place in a living organism.

Nanometre. One thousandth millionth of a metre (10^{-9} m).

Narrow-band Analysis. Analysis into single frequencies or a small range on either side of a specific frequency.

Neural Noise. Random nerve-cell activity, which increases with age and reduces visual discrimination.

Neuron. The basic unit of the nervous system, the neuron consists of a nerve cell with fibres which bring impulses in (dendrites) and fibres which conduct them out (axons).

Newton (N). The unit of force in the SI system, defined as that force which, applied to a free body with a mass of one kilogram, would give it an acceleration of one metre per second per second.

Non-destructive Testing. Test methods which do not result in destroying the test specimens.

Octave. The interval between two tones, one of which has twice the frequency of the other.

Octave-band Analysis. The division of a spectrum of noise into frequency ranges of octave intervals and the measurement of the sound intensities for the frequencies within them.

Oscillograph. An oscilloscope with a photographic system to record the waveforms, etc., displayed.

Oscilloscope. A cathode-ray tube arranged to trace waveforms and other patterns by deflection of a beam of electrons.

Permeable. Term used of a textile through which air and moisture vapour percolate. The problem of condensation beneath a waterproof garment is avoided by using a permeable material.

Pilomotor. Causing movements of hair. A goose pimple erects a hair on the skin.

Proprioceptors. Internal receptors which monitor position, velocity and acceleration of limbs.

Psychological Set. A temporary orientation, expectation or state of readiness to respond in a particular way to a particular stimulus.

Psychic Stress. Emotion, anxiety, conflict, frustration, worry about personal problems or arousal.

Pulmonary Ventilation. The volume of air pumped by the lungs, usually expressed in litres per minute.

Read-out. A method of presenting the results of measurement in letters and digits.

Root Mean Square (RMS) Value. The effective value of a fluctuating quantity. The values of the quantity are squared and averaged; then the square root of this average is extracted.

Saccadic Movements. Jerky movements of the eyes when searching a field of view.

SI Units. The International System of Units (*'Système international d'unites'*). There are seven base units: metre (length), kilogram (mass), second (time), ampère (electric current), Kelvin (temperature), candela (luminous intensity) and mole (amount of substance).

Sound Analyser. A sound-level meter combined with a frequency analyser arranged to measure sound pressures over specific bands of frequencies.

Spectrum. An arrangement of frequencies in order of those frequencies, from low to high.

Specular Surface. A shiny, as opposed to matt, surface; examples are mirrors and polished metals.

Stereo Vision. The ability to judge distance in a visual field.

Subliminal Stimulus. Stimulus below the threshold of sensitivity. For example, pictures shown for too short a time to be consciously apprehended can produce effects on the viewer.

Synapse. The mode of connexion of one nerve cell with another, a gap across which impulses are transmitted in one direction by chemical action. The chemicals concerned are termed transmitter chemicals.

Threshold. The minimum intensity of a stimulus required to trigger off a nerve impulse.

Transducer. A device to convert one form of energy to another. Examples are the microphone converting acoustic power to electrical power and the electric motor converting electrical power to mechanical power.

Wave. A disturbance propagated in a medium. Sound waves are fluctuations of pressure in an elastic medium (e.g., air). Light and heat waves occur in the electromagnetic spectrum; the disturbances are changes in the electric and magnetic fields.

Wavefront. A surface at all of whose points the phase is the same.

Wavelength. The distance between two similar and successive points on an alternating wave; for example, between successive maxima or successive minima. It is also the perpendicular distance between two wavefronts differing in phase by one complete period. It is equal to the velocity of proprgation of the wave divided by the frequency of the alternations.

Index